Sustainable Energy Policies and Technology

WIT_PRESS_

WIT Press publishes leading books in Science and Technology.
Visit our website for the current list of titles.
www.witpress.com

WIT_eLibrary_

Home of the Transactions of the Wessex Institute.
The WIT electronic-library provides the international scientific community with immediate
and permanent access to individual papers presented at WIT conferences.
http://library.witpress.com

Sustainable Energy Policies and Technology

Editor

S. Syngellakis
Wessex Institute, UK

WITPRESS Southampton, Boston

Editor:

S. Syngellakis
Wessex Institute, UK

Published by

WIT Press
Ashurst Lodge, Ashurst, Southampton, SO40 7AA, UK
Tel: 44 (0) 238 029 3223; Fax: 44 (0) 238 029 2853
E-Mail: witpress@witpress.com
http://www.witpress.com

For USA, Canada and Mexico

Computational Mechanics International Inc
25 Bridge Street, Billerica, MA 01821, USA
Tel: 978 667 5841; Fax: 978 667 7582
E-Mail: infousa@witpress.com
http://www.witpress.com

British Library Cataloguing-in-Publication Data
A Catalogue record for this book is available
from the British Library

Library of Congress Catalog Card Number: 2021950055

ISBN: 978-1-78466-455-8
eISBN: 978-1-78466-456-5

The texts of the papers in this volume were set individually by the authors or under their supervision. Only minor corrections to the text may have been carried out by the publisher.

Preface

Energy policies and management are of primary importance to achieve the development of sustainability and need to be consistent with recent advances in energy production and distribution. Progressing from an economy mainly focussed on hydrocarbons to one taking advantage of sustainable renewable energy resources requires considerable scientific research, as well as the development of new engineering systems.

Energy fuels the world's economy. Diminishing resources and severe environmental effects resulting from the continuous use of fossil fuels has motivated an increasing amount of interest in renewable energy resources and the search for sustainable energy policies.

Key difficulties to overcome often originate from the conversion of renewable energies (wind, solar, etc.) to useful forms (electricity, heat, fuel) at an acceptable cost, including impacts on the environment as well as in the integration of these resources into the existing infrastructure.

A wide range of topics are covered by the works contained in this book. The collaboration of varied disciplines are involved in order to arrive at optimum solutions, including studies of materials, energy networks, new energy resources, storage solutions, waste to energy systems, smart grids and many others.

The Editor wishes to acknowledge the authors, the members of the Scientific Committee, the Referees, the Institutional Partners, who supported the Energy and Sustainability 2021 conference

The Editor, 2022

Contents

THERMAL PERFORMANCE INVESTIGATION OF A MINI NATURAL CIRCULATION LOOP FOR SOLAR PV PANEL OR ELECTRONIC COOLING SIMULATED BY LATTICE BOLTZMANN METHOD

JOHAN AUGUSTO BOCANEGRA, ANNALISA MARCHITTO & MARIO MISALE
DIME – Department of Mechanics, Energetics, Management, and Transportation, Thermal Engineering and Environmental Conditioning Division, University of Genoa.

ABSTRACT

The natural circulation loop (NCL) consists of a thermal-hydraulic system that convoys thermal energy from a heat source to a heat sink without a pump. Applications of those loops can be found in solar energy, geothermal, nuclear reactors, and electronic cooling. The lattice Boltzmann method is a numerical method that can simulate thermal-fluid dynamics, using a mesoscopic approach based on the Boltzmann equation for the density function. A square NCL model with fixed temperatures at the heater and heat sink sections was developed in a bi-dimensional lattice with double distribution dynamics, one distribution for the hydrodynamic field and the other for the thermal field. The different cooler–heater configurations (vertical or horizontal) were investigated. We found that by positioning the source or sink vertically, the flow direction can be controlled. In contrast, in a loop with symmetric horizontal heater - horizontal cooler configuration where both fluid directions are equally probable. The effectiveness of the loop was studied by calculating the heat sink temperature gradient. The lower value was obtained for the horizontal heater horizontal cooler orientation (0.71) and the higher value for the vertical heater vertical cooler configuration with an increment of 34%; simultaneously, the flow rate (Reynolds number) was reduced by 47%.
Keywords: end heat exchanger, heater orientation, heat sink orientation, LBM, thermosyphon.

1 INTRODUCTION

Thermal energy is dissipated from several systems, such as electronics, computers, and mechanical machinery. This dissipated energy can be considered as waste heat. On the other hand, several energy production systems use heat as the primary source, e.g. geothermic generators or nuclear power plants. The heat involved in those systems can be used *efficiently* if optimized technology is used to transport this heat from the source to the desired heat sink. This *efficiency* must consider different thermodynamic factors (as heat transfer effectiveness), environmental factors (as the use of non-toxic substances and reduced energy consumption), and economic factors (as implementation cost and maintenance costs).

Among the systems that need an effective cooling system in a mini-scale (under-meter systems) are common electronics and solar collectors. For example, in electronics and computation, the new trend of mining cryptocurrencies [1, 2] and the use of massive computation clusters and data centers [3] constantly emit heat that must be efficiently transported to a convenient sink (in many applications can be simply the ambient). On the other hand, a very interesting application can be found in the solar energy field: the PV panels work better at low temperatures, then the solar energy can be profited implementing thermal-voltage hybrid systems (PV-T), the heat can be stored and used for water heating [4], or can be used directly for other applications as dryer systems [5].

© 2022 WIT Press, www.witpress.com
DOI: 10.2495/EQ-V7-N1-1-12

Natural circulation (or free convection) is of great interest in the energetic and sustainability context because it is possible to convey the thermal energy without using pumps, using closed natural circulation loops (NCLs). Different typologies of NCLs exist. For example, these can be classified by the working fluid typology as single-phase or two-phase NCL. Also, these can be categorized regarding their geometry: the most common geometries are rectangular-shaped or circular-shaped. Some other complex configurations exist as toroidal theta loops [6], parallel-coupled loops [7], or series-coupled loops [8]. The position of the heater and the cooler gives origin to several loop configurations named as horizontal heater horizontal cooler (HHHC), horizontal heater vertical cooler (HHVC), vertical heater horizontal cooler (VHHC), and vertical heater vertical cooler (VHVC) [9]. The use of NCL has, in addition to the mentioned absence of a pump, the great advantage to using different working fluids, like air, water, CO_2, dielectric fluids, or nanofluids [10, 11], and is common to work with non-toxic materials like water or air as working fluids [12, 13].

This study presents results considering a case study of a (mini) NCL suitable for electronic cooling or solar PV panel cooling. The thermohydraulic performance of the considered systems was simulated by using the lattice Boltzmann method (LBM). This study shows the effects of changing the loop configuration (HHHC, HHVC, VHHC, VHVC) on the flow characteristics and the thermal effectiveness. The HHHC configuration presents a higher flow rate for the same operational conditions.

2 METHODOLOGY

2.1 NCL description

Figure 1 presents a possible application of NCL. This study focus on a single mini-loop with a rectangular shape. The diameter is small, and this selection is not trivial because it is known that stable behavior is achieved if small inner diameters are considered or localized pressure losses are implemented along the circuit [14]. The working fluid is water.

The comparison among the four configurations was made at the same operating conditions in the laminar regime. Those operating conditions can be resumed in the Rayleigh number; in this case, the value 10^6 was selected. A schematic representation of the loop and the tested configurations is presented in Fig. 2. The loop considered here works with the fixed temperature at the heater (T_H) and cooler (T_C), also known as *NCL with heat end exchangers* [15]. Recently was shown that this kind of square mini-loops can work connected in parallel to convoy more thermal energy, and the thermohydraulic performance remains stable even if the heating power in one circuit changes [7, 16].

2.2 LBM for thermal flow

The LBM is a numerical method used to simulate fluids by solving the discretized Boltzmann equation (LBE) eqn (1). This relationship expresses the balance between transport and collision processes when a statistical approach over the molecules of the fluid is considered. Complex boundaries and vorticity can be simulated by this numerical method [17]. A review of the application of this method in nuclear reactor problems, including thermal flows and neutronics, can be found in Bocanegra *et al.* [18].

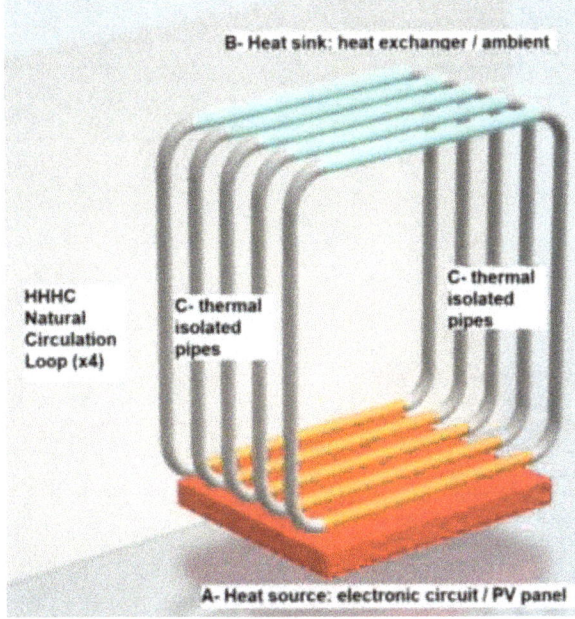

Figure 1: Application example: four NCLs in HHHC configuration transport the heat from A (circuit or PV panel, hot temperature) to B (heat sink, cold temperature). The number of loops can vary for each application.

$$f_i\left(\mathbf{r}+\mathbf{c}_i\Delta t,t+\Delta t\right)-f_i\left(\mathbf{r},t\right)=-\frac{1}{\tau}\left(f_i\left(\mathbf{r},t\right)-f_i^{eq}\left(\mathbf{r},t\right)\right)+\Delta t\ F_i \qquad (1)$$

The evolution of the probability density function f_i is described by the LBE. The probability density function f_i represents the probability of finding a molecule with a given velocity in each position $\mathbf{r}$ and time t. This probabilistic approach is possible considering a discretization of space $\Delta\mathbf{r}$, time Δt, and velocity $\mathbf{c}_i$. In this way, a lattice structure propagates the density function in given directions characterized by the discretized velocity set $\mathbf{c}_i$. The lattice structure used for the simulation is known as D2Q9. To simulate the temperature field and additional distribution function, g_i was used with a D2Q5 lattice structure, eqn (2). The double distribution function approach has been applied successfully to diverse thermal fluid dynamics problems, e.g. natural circulation in cavities or heat transfer in a microchannel [19]. The two-way coupling between the hydraulic field and the thermal field was achieved by considering a forcing term F_i proportional to the local temperature representing the buoyancy force (Boussinesq hypothesis). The approach adopted is similar to the one presented by Guo *et al.* [20]

$$g_i\left(\mathbf{r}+\mathbf{c}_i\Delta t,t+\Delta t\right)-g_i\left(\mathbf{r},t\right)=-\frac{1}{\tau_g}\left(g_i\left(\mathbf{r},t\right)-g_i^{eq}\left(\mathbf{r},t\right)\right) \qquad (2)$$

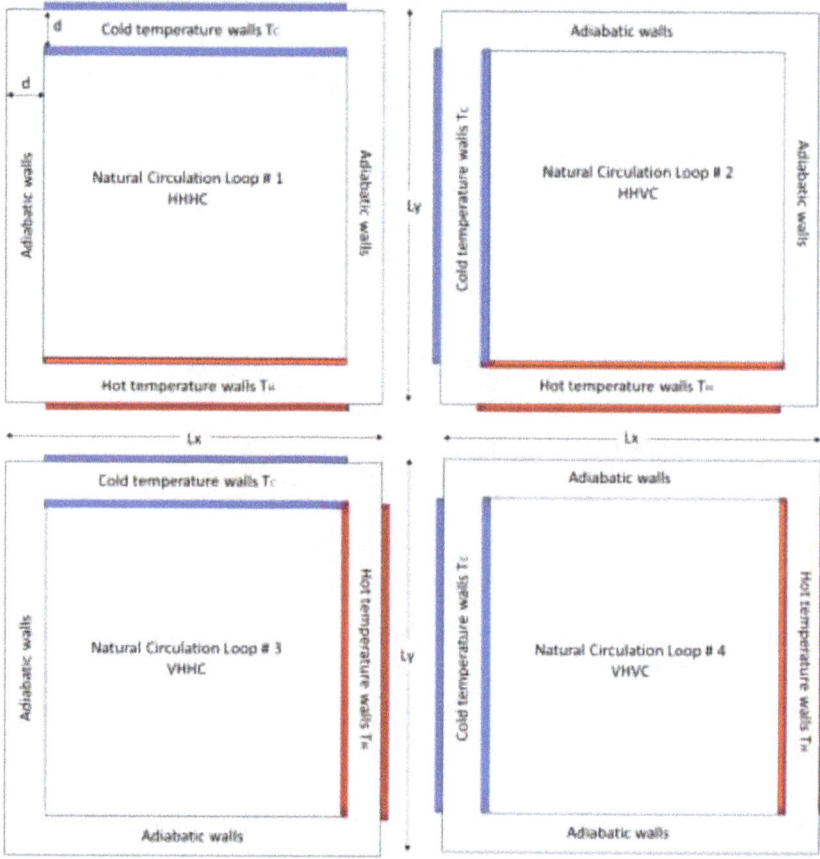

Figure 2: Loop configurations: horizontal heater horizontal cooler (HHHC), horizontal heater vertical cooler (HHVC), vertical heater horizontal cooler (VHHC), and vertical heater vertical cooler (VHVC). $L_x = L_y = 0.250$ m, pipe diameter $d = 0.010$ m.

The equilibrium distributions f_i^{eq} and g_i^{eq} are calculated as a Taylor expansion of the Maxwell–Boltzmann distribution using the velocity and truncated holding second-order terms. τ, τ_g represent a time scale for the relaxation to equilibrium process the kernel of the collision algorithm presented on the right-hand side of eqns (1) and (2). τ is proportional to the viscosity γ. The ratio of these two relaxation times is equivalent to the Prandtl number (Pr) of the simulated fluid; in our case, we use Pr=7.0. More details of the LBM can be found in [21–23].

Some advantages of the LBM are linked to the parallelization capabilities by the application of local collision and streaming rules in each iteration, and the easy handle of complex geometry using the lattice structure instead of the typical meshing algorithms used in common CFD methods.

The code used for the simulations was implemented in C++ using the PALABOS library [24]. Message passing interface (MPI) standard library was used for the parallel implementation and run of the code on an Intel® Xeon® Platinum 8260 CPU, 2.40 GHz, workstation using 46 cores. The average velocity for the runs is 157 Mega site updates per second (MSUPS).

3 RESULTS AND DISCUSSION

3.1 Model validation

The numerical model was validated by contrast with the analytical and empirical model proposed by Cheng *et al.* [25] for a single-phase NCL in HCHH configuration with fixed temperature boundaries. All the expressions used for validation were previously verified experimentally by Cheng *et al.* [25]. The numerical results were compared with the physical values using non-dimensional groups and normalized variables. The temperature was normalized using 0 for the heat sink and 1 for the heater. The velocity was normalized using the maximum steady-state velocity along the loop. The steady-state Reynolds number ($Re_{ss} = v_{rms}^{ss} d / \gamma$) is referenced to the pipe diameter d and is proportional to the rms velocity (and flow rate), e.g. a rms velocity of 0.001 m/s is equivalent to a Re_{ss}=10, approximately. Accordance between the analytical model and the LBM results for the thermal field was observed, details are provided in Fig. 7a. In the same manner were contrasted the analytical equation to determine the steady-state Reynolds number (error below 5%) and the empirical relationship for the Nusselt number (error below 15%). Additionally, the performance of the boundary conditions was evaluated by contrasting the parabolic velocity cross-section profile into the pipes with a very good fitting and confirm the no-slip boundary condition imposed. The thermal boundary conditions were verified by observing the temperature gradient near the adiabatic walls, and near-zero values were found. After that validation, the model was modified to study the effect of locating the cooler and heater in vertical positions. The detailed contrast with the analytical model under different operation conditions is object of a future work.

3.2 Fluid velocity profile

As was expected, the HHHC loop is the only symmetrical configuration considered, and the flow in both senses (clockwise and counterclockwise) is possible. We observe this fact under several runs of the code. On the other hand, the vertical position of the cooler or the heater forces the flow in the loop in the considered geometry in a counterclockwise sense. Figure 3 shows the flow under the four configurations analyzed here.

In Fig. 3, the flow pattern is laminar in all the cases. The velocity decreases to zero near the walls and presents the maximum in the center of the tube. However, some small perturbances are noted at the corners, probably by the 90° elbows. Moreover, a big deviation for the expected parabolic profile was noted at the inlet of the vertical heaters or coolers. This effect can be explained by the buoyancy force acting against the fluid flow in those vertical pipes. This interesting fact is depicted in Fig. 4a; the cross-section velocity profile at different locations along the vertical heater shows a sensible decrease of the velocity at the center near the inlet (three diameters), and then the effect vanishes, and the parabolic profile is developed. Figure 4b depicts the longitudinal velocity profile for the VHVC loop to illustrate this phenomenon at the center of the pipe. However, the effect on the cross-section averaged velocity is almost imperceptible.

Figure 5 shows the transient behavior of the four circuits. The oscillation during the transient is more evident for the HHHC configuration, similar for HHVC and VHHC, and very reduced for VHVC. Also, it is visible in the figure that the time to arrive at the steady state changes for each configuration. Similar behaviors were experimentally observed by Vijayan

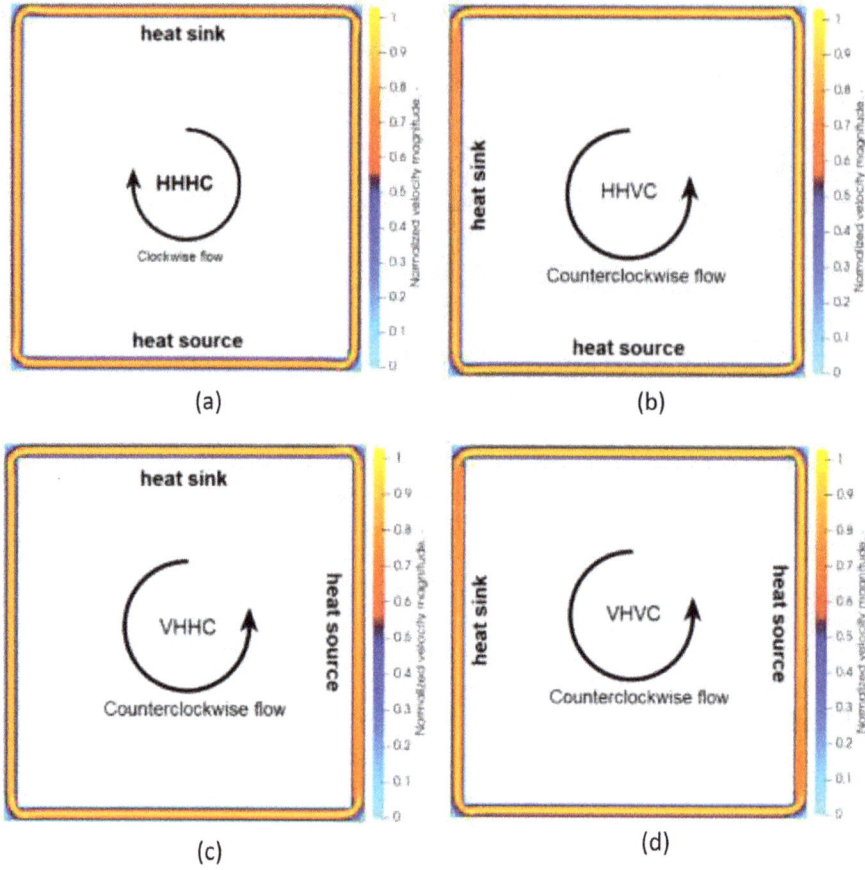

Figure 3: Normalized velocity magnitude and flow direction along the loops at steady-state (a) HHHC; (b) HHVC; (c) VHHC; (d) VHVC.

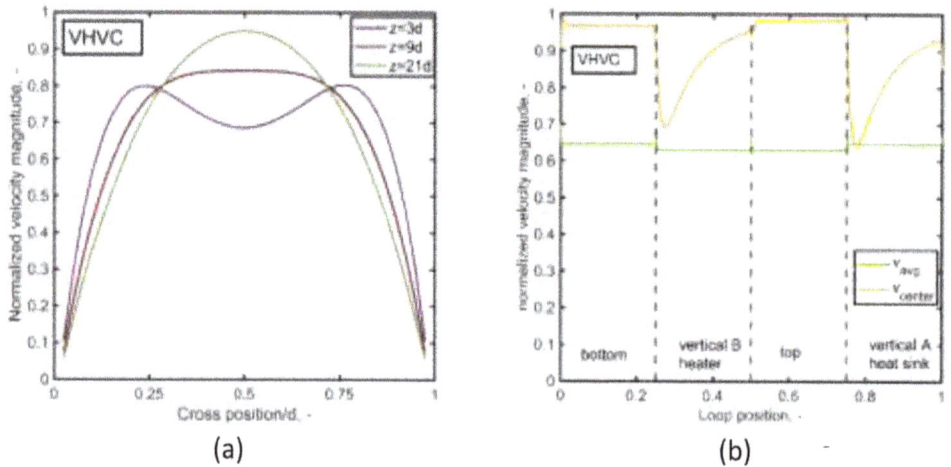

Figure 4: VHVC velocity profile: (a) Cross-section and (b) longitudinal section at the center of the pipe v_{center} and cross-section average velocity v_{avg}.

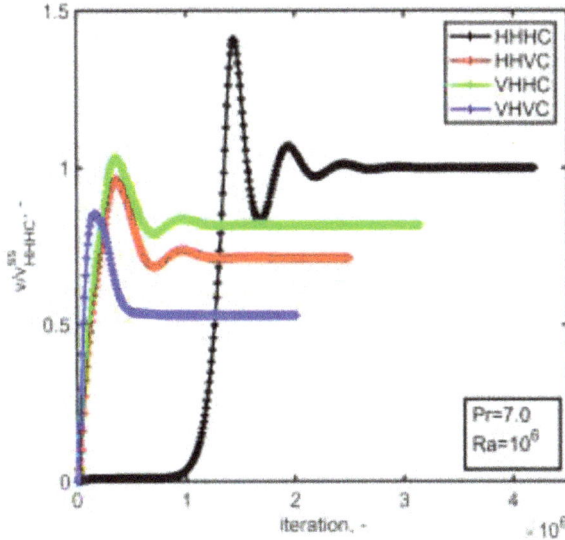

Figure 5: Transient of the velocity, normalized to the HHHC velocity at steady state.

et al. [26] and Chen *et al.* [27]. They found experimentally that for HHHC and HHVC, the time to start the circulation is higher than the time to start the flow with vertical heaters, in which cases the flows start when the heating power is provided.

3.3 The thermal field at steady state

Figure 6 shows the thermal field at the steady state. It is notorious that the main temperature gradient is obtained at the heaters and heat sinks. Moreover, the other pipes present a minimum temperature variation. In fact, this figure provides a visual probe of the adiabatic condition of those unheated pipes. Also, it is interesting to note that in the HHHC, one adiabatic pipe is hot, and the other is cold, as in the VHVC configuration. On the other hand, the HHVC and VHHC configurations have only one adiabatic section, hot and cold, respectively. Figure 7 presents the normalized temperature profile along the longitudinal axe.

3.4 Thermohydraulic performance

To compare the performance of the four circuits, some parameters and non-dimensional groups were calculated (Table 1). The percentual differences referenced to the HHHC values are included for the HHVC, VHHC, and VHVC loops.

Table 1 shows the simulation steps needed to reach the steady state. The Reynolds number referenced to the pipe diameter is used to evaluate the flow regime. It is noted that the highest value is obtained in the HHHC configuration. The Nusselt number expresses the proportion between the convective and conductive heat transfer. The average Nusselt number was numerically calculated for the heater in all the configurations using the temperature gradient near the wall. The theoretical Nusselt number for the HHHC configuration was calculated by fitting the thermal profile to the Cheng *et al.* model (Nu = 1.60) and using the empirical correlation proposed by them (Nu = 1.43). This value is lower than the common value for straight

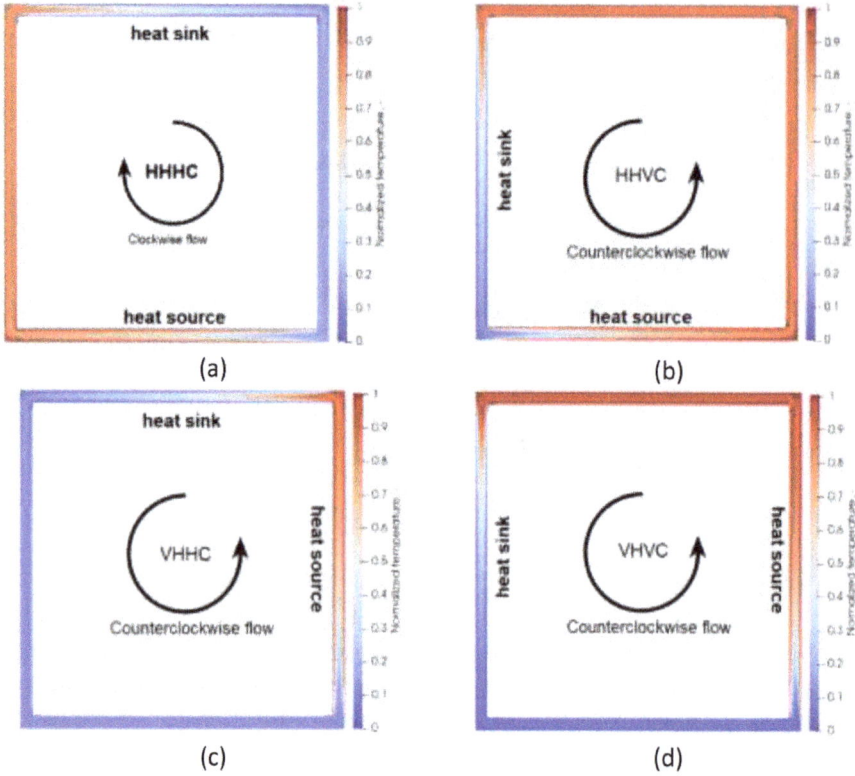

Figure 6: Thermal field and flow direction along the loops at steady state. (a) HHHC, (b) HHVC, (c) VHHC, (d) VHVC.

pipes $Nu = 3.66$. The highest value was observed for the HHHC configuration. The temperature difference at the heat sink inlet and outlet is included $T_{sink}^{in} - T_{sink}^{out}$, the highest value was obtained for the VHVC configuration. Using the temperature values at the inlet T_{sink}^{in} and T_{sink}^{out} outlet of the heat sink, it is possible to obtain the effectiveness, eqn (3), by comparing the steady-state temperature difference $T_{sink}^{in} - T_{sink}^{out}$, with the maximum possible difference $T_{sink}^{in} - T_C$. The highest heat sink effectiveness was found in the VHVC configuration.

$$\varepsilon_{sink} = \frac{T_{sink}^{in} - T_{sink}^{out}}{T_{sink}^{in} - T_C} \tag{3}$$

4 CONCLUSIONS

A bi-dimensional model of a single-phase NCL with heat end exchangers was simulated using the LBM with double distribution functions. The model was validated using the model proposed by Cheng *et al.* [25] equivalent to Vijayan's laminar regime model [12] for the Horizontal Heater Horizontal Cooler configuration. The effect of considering different configurations for the loop was presented for the transient and the steady state. Moreover, the observations agree with previous observations presented by Vijayan *et al.* [26].

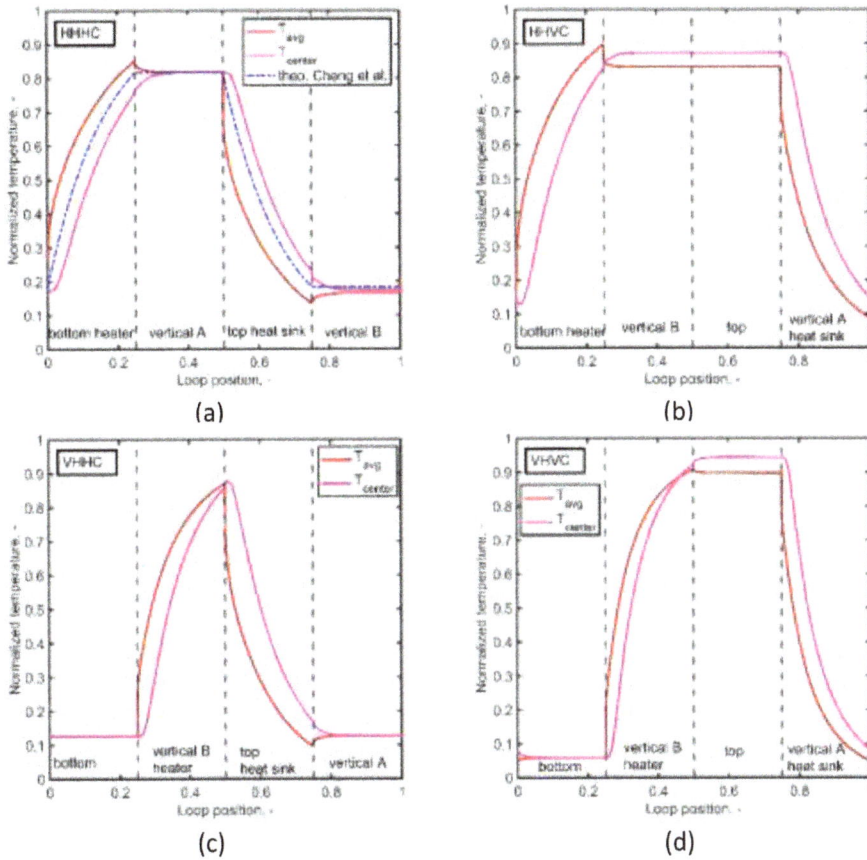

Figure 7: Temperature profile at steady state along the loop (following the flow direction). (a) HHHC, (b) HHVC, (c) VIIIC, (d) VIIVC.

- The HHHC configuration gives the highest flow rate.
- Disturbances in the parabolic velocity profile were observed at the inlet of the vertical section that contains the heater or cooler.
- The natural circulation takes more time to take place in the HHHC configuration, and this time is similar for the other three configurations.
- The time to reach the steady state going from low to high is VHVC, HHHC, VHHC, and HHHC, respectively.
- The VHVC configuration reached higher heat sink effectiveness.
- The highest Nusselt number was obtained in the HHHC configuration.

Finally, this study shows that this kind of square loop is suitable for cooling systems of small components (as electronics or solar PV panels) in all the studied configurations. Under the tested operational parameters, all the loops remain in the laminar regime; the dynamic was stable (no flux inversion), and all reached a steady state. However, the thermohydraulic performance depends strongly on the selected configuration. The two extreme cases were the

Table 1: Thermohydraulic parameters were evaluated at steady state, considering the four heater–cooler configurations. The comparison took the HHHC values as a reference.

Non-dimensional parameter	Loop configuration			
	HHHC	HHVC	VHHC	VHVC
Transient time, timesteps	4,195,300	2,494,300	3,143,900	2,013,000
		–41%	–25%	–52%
Reynolds number:	15.2	10.9	12.5	8.1
		–28%	–18%	–47%
Heater Nusselt number:	1.78	0.43	0.43	0.37
		–76%	–76%	–79%
Heat sink temperature difference:	0.58	0.74	0.70	0.85
		+28%	+21%	+47%
Heat sink effectiveness:	0.71	0.89	0.80	0.94
		+25%	+13%	+32%

HHHC (with the highest Reynolds number at steady state and lower sink effectiveness) and the VHVC (with the lowest Reynolds number at steady state and higher sink effectiveness). On the other hand, the performance of the HHVC and VHHC is similar. Future research must be done concerning the variation of the thermohydraulic performance at different temperature gaps, i.e. varying the Rayleigh number.

ACKNOWLEDGMENT

This research was funded by Ministero dell'Istruzione, dell'Universita e della Ricerca (MIUR, Italy), grant number PRIN-2017F7KZWS.

REFERENCES

[1] Náñez Alonso, S.L., Jorge-Vázquez, J., Echarte Fernández, M.Á. & Reier Forradellas, R.F., Cryptocurrency mining from an economic and environmental perspective. Analysis of the most and least sustainable countries. *Energies*, **14**, pp. 4254, 2021. https://doi.org/10.3390/en14144254

[2] de Vries, A., Bitcoin's growing energy problem. *Joule,* **2**, pp. 801–805, 2018. https://doi.org/10.1016/j.joule.2018.04.016

[3] Huang, P., Copertaro, B., Zhang, X., Shen, J., Löfgren, I., Rönnelid, M., Fahlen, J., Andersson, D. & Svanfeldt, M., A review of data centers as prosumers in district energy systems: renewable energy integration and waste heat reuse for district heating. *Applied Energy*, **258**, pp. 114109, 2020. https://doi.org/10.1016/j.apenergy.2019.114109

[4] Ju, X., Xu, C., Liao, Z., Du, X., Wei, G., Wang, Z. & Yang, Y., A review of concentrated photovoltaic-thermal (CPVT) hybrid solar systems with waste heat recovery (WHR). *Science Bulletin*, **62**, pp. 1388–1426, 2017. https://doi.org/10.1016/j.scib.2017.10.002

[5] Al-Kayiem, H.H., Hybrid techniques to enhance solar thermal: the way forward. *Int. J. EQ.*, **1**, pp. 50–60, 2015. https://doi.org/10.2495/EQ-V1-N1-50-60

[6] Satou, A. Madarame, H. & Okamoto, K., Unstable behavior of single-phase natural circulation under closed loop with connecting tube. *Experimental Thermal and Fluid Science*, **7**, 2001.

[7] Misale, M., Bocanegra, J.A., Borelli, D. & Marchitto, A., Experimental analysis of four parallel single-phase natural circulation loops with small inner diameter. *Applied Thermal Engineering*, **180**, pp. 115739, 2020. https://doi.org/10.1016/j.applthermaleng.2020.115739

[8] Dass, A. & Gedupudi, S., Numerical investigation on the heat transfer coefficient jump in tilted single-phase natural circulation loop and coupled natural circulation loop. *International Communications in Heat and Mass Transfer*, **120**, pp. 104920, 2021. https://doi.org/10.1016/j.icheatmasstransfer.2020.104920

[9] Misale, M., Overview on single-phase natural circulation loops, in: International Conference on Advances in: Mechanical & Automation Engineering, Rome, Italy, 2014: p. 13.

[10] Misale, M., Devia, F. & Garibaldi, P., Experiments with Al2O3 nanofluid in a single-phase natural circulation mini-loop: Preliminary results. *Applied Thermal Engineering*, **40**, pp. 64–70, 2012. https://doi.org/10.1016/j.applthermaleng.2012.01.053

[11] Garibaldi, P. & Misale, M., Experiments in single-phase natural circulation miniloops with different working fluids and geometries. *J. Heat Transfer*, **130**, pp. 104506, 2008. https://doi.org/10.1115/1.2948393

[12] Vijayan, P.K., Experimental observations on the general trends of the steady state and stability behaviour of single-phase natural circulation loops. *Nuclear Engineering and Design*, **215**, pp. 139–152, 2002. https://doi.org/10.1016/S0029-5493(02)00047-X

[13] Misale, M., Experimental study on the influence of power steps on the thermohydraulic behavior of a natural circulation loop. *International Journal of Heat and Mass Transfer*, **99**, pp. 782–791, 2016. https://doi.org/10.1016/j.ijheatmasstransfer.2016.04.036

[14] Misale, M. & Frogheri, M., Influence of pressure drops on the behavior of a single-phase natural circulation loop: preliminary results. *International Communications in Heat and Mass Transfer*, **26**, pp. 597–606, 1999. https://doi.org/10.1016/S0735-1933(99)00046-9

[15] Cheng, H., Lei, H. & Dai, C., Thermo-hydraulic characteristics and second-law analysis of a single-phase natural circulation loop with end heat exchangers. *International Journal of Thermal Sciences*, **129**, pp. 375–384, 2018. https://doi.org/10.1016/j.ijthermalsci.2018.03.026

[16] Misale, M., Bocanegra, J.A. & Marchitto, A., Thermo-hydraulic performance of connected single-phase natural circulation loops characterized by two different inner diameters. *International Communications in Heat and Mass Transfer*, **125**, p. 105309, 2021. https://doi.org/10.1016/j.icheatmasstransfer.2021.105309

[17] Arumuga Perumal, D. & Dass, A.K., Simulation of flow in two-sided lid-driven square cavities by the lattice Boltzmann method, in: The New Forest, UK, 2008: pp. 45–54. https://doi.org/10.2495/AFM080051

[18] Bocanegra Cifuentes, J.A., Borelli, D., Cammi, A., Lomonaco, G. & Misale, M., Lattice Boltzmann method applied to nuclear reactors—a systematic literature review. *Sustainability*, **12**, p. 7835, 2020. https://doi.org/10.3390/su12187835

[19] Sousa, A.C.M., Hadavand, M. & Nabovati, A., Three-dimensional simulation of slip-flow and heat transfer in a microchannel using the lattice Boltzmann method, in: Tallinn, Estonia, pp. 75–85, 2010. https://doi.org/10.2495/HT100071

[20] Guo, Z., Shi, B. & Zheng, C., A coupled lattice BGK model for the Boussinesq equations. *Int. J. Numer. Meth. Fluids,* **39**, pp. 325–342, 2002. https://doi.org/10.1002/fld.337

[21] Succi, S., The Lattice Boltzmann *Equation for Fluid Dynamics and Beyond,* Clarendon Press, Oxford, England, 2001.

[22] Succi, S., Benzi, R. & Massaioli, F., A review of the Lattice Boltzmann method. *International Journal of Modern Physics C,* **4**, pp. 409–415, 1993.

[23] Arumuga Perumal, D. & Dass, A.K., A review on the development of lattice Boltzmann computation of macro fluid flows and heat transfer. *Alexandria Engineering Journal,* **54**, pp. 955–971, 2015. https://doi.org/10.1016/j.aej.2015.07.015

[24] Latt, J., Malaspinas, O., Kontaxakis, D., Parmigiani, A., Lagrava, D., Brogi, F., Belgacem, M.B., Thorimbert, Y., Leclaire, S., Li, S., Marson, F., Lemus, J., Kotsalos, C., Conradin, R., Coreixas, C., Petkantchin, R., Raynaud, F., Beny, J. & Chopard, B., Palabos: Parallel Lattice Boltzmann Solver. *Computers & Mathematics with Applications,* p. S0898122120301267, 2020. https://doi.org/10.1016/j.camwa.2020.03.022

[25] Cheng, H., Lei, H., Zeng, L. & Dai, C., Theoretical and experimental studies of heat transfer characteristics of a single-phase natural circulation mini-loop with end heat exchangers. *International Journal of Heat and Mass Transfer,* **128**, pp. 208–216, 2019. https://doi.org/10.1016/j.ijheatmasstransfer.2018.08.136

[26] Vijayan, P.K., Sharma, M. & Saha, D., Steady state and stability characteristics of single-phase natural circulation in a rectangular loop with different heater and cooler orientations. *Experimental Thermal and Fluid Science,* **31**, pp. 925–945, 2007. https://doi.org/10.1016/j.expthermflusci.2006.10.003

[27] Chen, L., Zhang, X.-R. & Jiang, B., Effects of heater orientations on the natural circulation and heat transfer in a supercritical CO2 rectangular loop. *Journal of Heat Transfer,* **136,** p. 052501, 2014. https://doi.org/10.1115/1.4025543

ZERO-ENERGY BUILDINGS IN CITIES WITH DIFFERENT CLIMATES AND URBAN DENSITIES: ENERGY DEMAND, RENEWABLE ENERGY HARVEST ON-SITE AND OFF-SITE AND TOTAL LAND USE FOR DIFFERENT RENEWABLE TECHNOLOGIES

UDO DIETRICH
REAP Research Group (Resource Efficiency in Architecture and Planning),
HafenCity University Hamburg, Germany.

ABSTRACT

Zero-energy buildings (ZEBs) have no fossil energy consumption; this is achieved by optimizing the building and balancing the remaining energy needs by renewables. If this energy can be harvested on-site, on the building's envelope and its estate, a net-ZEB is reached. If supplementary renewable energy has to be produced off-site on compensating land, the ZEB can be reached with such compensating measures (ZEB_CM). Climate and urban density determine how far a ZEB is possible. Temperatures out of comfort range, lack of daylight and overheating by solar radiation may cause energy demand while high insolation or wind speed delivers good preconditions to produce renewable energy on less land. A high urban density avoids urban sprawl and saves land outside of the cities that can be used for other purposes (agriculture and energy production, among others). But, at a certain density, net-ZEB cannot be realized furthermore, and compensating land is necessary. The paper investigates these effects for 15 selected cities around the globe, covering all main climatic conditions. Based on design rules out of literature and own experiences, a prototypical optimized building is derived for each location, and its energy demand is simulated. Standard assumptions for the efficiency of renewable energy systems are used to determine the need of land to cover it. For different urban densities, it can be concluded how far net-ZEB is possible; if necessary, the need for compensating land is calculated. The results show that for cities with moderate climates, the total land use (city plus compensating land) can decrease with increasing urban density if the technology used off-site has high efficiency (like PV). On the other hand, the total land use may increase remarkably with increasing urban density if the used technology off-site has a low efficiency (like the wind for electricity and especially wood pellets for heating). The final understanding is that cities should meet the energy needs on-site by optimized buildings and structures plus renewable energy production (PV on the building's roofs, geothermal systems, etc.).
Keywords: compensating measures, different climates, optimized buildings, urban density, zero-energy building.

1 INTRODUCTION

Buildings account for a considerable share of energy for their operation and maintenance leading to a significant impact on the environment. It is estimated that 30% of the global share of energy is consumed by commercial and residential buildings leading to 28% of global emissions (excluding construction industry) [1]. The demand for energy by buildings is expected to show an upward trend in the coming years. Residential and commercial buildings consume approximately 60% of the world's electricity [2]. Eighty-two percent of final energy consumption in buildings was supplied by fossil fuels in 2015 (including primary energy input for power generation; traditional use of biomass excluded) [3].

The Paris Agreement on climate change in the year 2015 charted a new course in an effort to check global warming. Sustainable measures towards energy-efficient and low-carbon solutions for buildings and construction can help achieve the central aim of this agreement, a carbon-free society in a few decades.

© 2022 WIT Press, www.witpress.com
DOI: 10.2495/EQ-V6-N4-335-346

The concept of zero-energy buildings (ZEBs) is required to reduce energy consumption and bring down CO_2 emission. The 'Zero' refers to the primary (=fossil) energy demand for the services that are necessary to guarantee the user's comfort inside:

– Heating
– Cooling
– Electricity for artificial light
– Electricity for mechanical ventilation (fans)
– Domestic hot water

That target is achieved by increasing the efficiency of the building and balance the energy needs by renewables that are produced either on-site or off-site.

Following the definition, it must be noted that users of a ZEB do not, by far, live carbon-neutral! The electricity for the use itself ('tenants' electricity': PC, server, gadgets, TV), the energy for transportation, nutrition, etc. are NOT included and may not be considered negligible.

ZEBs can be divided into two classes: a) net ZEBs and b) ZEBs with compensating measures (ZEB_CM). The difference arises from the location of renewable energy production. In case of net ZEBs, the energy demand is covered on-site, with systems on the building's envelope and/or on the ground of the own estate. In case of a ZEB_CM that energy production on-site is not sufficient to cover the demand, thus supplementary compensating land/alternative renewable energy sources outside of the own estate must be used.

For a net ZEB, it must be stated that there is a competition between the area of usage that has an energy demand and the size of the building's envelope and estate to produce energy to cover it. With an increasing number of storeys, it becomes more and more difficult to balance the demand (the roof area of the building is the main area for renewable energy systems, it does not increase with the increasing number of storeys, the estate remains the same, etc.). Thus, it can be expected that a net ZEB is only possible to a certain, limited number of storeys and urban density.

In the other case, when compensating measures are necessary, a ZEB_CM will use renewable technologies like PV modules, wind turbines or wood pellets for energy generation on the compensating land.

The energy demand of a building is mainly influenced by its location and design; the main aspects are climatic conditions, urban density, orientation and positioning of the building, number of storeys, construction mass, window to wall ratio, daylight access, natural ventilation strategy, shading system, air tightness, thermal insulation, etc.

A second decisive question is whether the building's users have the possibility to use the building adaptively by personally adaptable thermostats, operable windows, shading systems, light switches, etc. and if they have the chance to adapt themselves to different indoor temperatures with their clothing (no dress code).

This paper is based on the hypothesis that the better the building is climate and user adaptive, the lesser is its energy demand.

2 SCOPE OF WORK/AIM OF THE PAPER

In this study, 15 major cities around the globe having different climatic characteristics are chosen. Cities exhibiting similar climatic features are grouped to find similarities and differences. Reykjavik, Oslo and Hamburg are in colder locations, having temperatures less than 10°C for most of the months and demand heating. Chicago and Beijing have both cold and warm months and require both heating and cooling. Cairo and Delhi have a hot and dry climate with the need for cooling. Singapore, Dar es Salaam, Jakarta and Santo Domingo are hot and humid and need air-conditioning almost throughout the year. Sydney, Santiago, Mexico-City and Addis Ababa belong to a widely comfortable category.

For these cities, the following research questions are investigated:

- What is the maximum number of storeys, size of the estate and urban density for which it is possible to reach net ZEB?
- What is the need of land for compensating measures for different urban densities for ZEB_CM? What is the influence of different renewable energy technologies (like PV versus wind turbines, etc.)?

3 METHODOLOGY

This research sums up and generalizes findings based on a university's master course. Different groups of international students worked on the different cities having different climates to find the suitability of ZEBs in these cities. Both net ZEBs and ZEB_CM were analysed and design strategies were adopted.

To facilitate the task, the study is carried out with office buildings that are composed of standard office rooms. They have standard conditions in use and design that are easy to describe. Of course, a major portion of the built-up area comprises residential buildings; it also includes retails, schools, hospitals and industries. Residential buildings need less energy than office buildings; thus if the ZEB for office buildings could be achieved, then it could also be achieved for residences.

Furthermore, real urban situations were not regarded since this is too complicated. A quarter of (identical) office building is assumed, well reflecting the effect of reduced daylight access and increased electricity demand for artificial light if buildings are near together and shade each other. It was mentioned that such mono-use quarters are not sustainable. As the scope of the work is to obtain general tendencies and comparable results with similar assumptions for different cities with diverse climatic conditions, it is difficult to consider the real urban situation. Urban density is expressed as the plot ratio, which is the ratio between the area of usage and the area of estate.

3.1 Standard office room size and equipment

The standard office room is a unit of 168 m^2 with a depth of 14 m and a width of 12 m, supposed to provide working space for 12 people. This model size could be replicated throughout the whole width and height of the buildings (with supplementing stairs and elevators etc.). Each storey height/floor height in the building is 3.2 m. The time of usage of the building is 11 hours and 5 days a week which is from morning 7 am to evening 6 pm from Monday to Friday.

3.2 Renewable energy production

3.2.1 On-site

The thermal energy is obtained by a geothermal system (see section 3.2.3) that fills in maximum the whole estate. The coefficients of performance (COP) for cooling and heating are 2.5 and 3.5, respectively. The size of the system determines the possible maximal power of the heating or cooling system.

The electricity necessary for heat pumps, artificial lighting and ventilation is received by harvesting solar energy through PV panels. Polycrystalline PV placed flat on the roof is used for the analysis. It is assumed that a surplus can be delivered to the grid and a supplementary need could feed out of the grid, thus the area of the PV modules is determining the annual contribution to the electricity demand of the building.

For a net ZEB, the described renewable energy systems can deliver the whole energy demand. It is determined up to which number of storeys of the building and up to which building distance/urban density that is possible.

3.2.2 On compensating land

In case of the assumed number of storeys or the urban densities being higher than the threshold for a net ZEB, a lack of heat and/or electricity is caused. To balance it, compensating land is needed to accomplish the energy demand of the building. Compensating electricity demand is obtained from onshore or offshore wind turbines or PV modules and supplied to the office building.

Transporting thermal energy (heat) from its source of generation to the place of intended use is not preferred as it leads to a lot of energy losses during its transport. Hence, transport of raw material on-site to burn there for heat energy is more economical. Wood pellets are chosen as a compensating measure for thermal energy demand.

Therefore, the area of compensating land depends on the annual energy demand as well as on the chosen type of renewable technology.

3.2.3 The energy density of different renewable energy systems

Solar radiation delivers – depending on the location – a few hundred to more than 2000 kWh/y energy to a square meter of earth's surface. Renewable energy systems transfer a part of it into usable energy (electricity, heat, material to burn). Figure 1 illustrates that the efficiency of this process differs remarkably from system to system (values based on [4]). Geothermal systems have highest efficiency (with distance) for heating and cooling and PV modules for electricity. Wind turbines onshore or offshore occupy a greater amount of land. But it must be noted that in some cases, it is difficult to have any other use (like agriculture) apart from PV modules in the land that is covered with PV as the land under PVs is dark and dry. Agri-PV is an option but, in any case, there are two systems, plants and PV, in competition for solar energy. The land in between the wind turbines in a wind farm could be used effectively for other purposes.

Energy plants hold the lowest energy density. Energy plants create competition to agriculture for food production and it should be avoided. Wood pellets are a by-product of renewable forestry; besides the limited production, there is no negative impact.

These values give the yearly harvest of renewable energy. The solar offer of 953 kWh/m² y and the harvest of PV modules (efficiency about 15%) refer to the location of Hamburg, Germany. For all other cities under investigation in this article, these values are adapted to the corresponding solar offer at the chosen location.

Data for wind refer to an average wind velocity of 3 m/s on land and 5 m/s on the sea and are assumed the same for all locations.

Own estimations show that the harvest of geothermal systems varies only slightly with the temperature in the ground and thus the location. The system is assumed as 100 m deep vertical

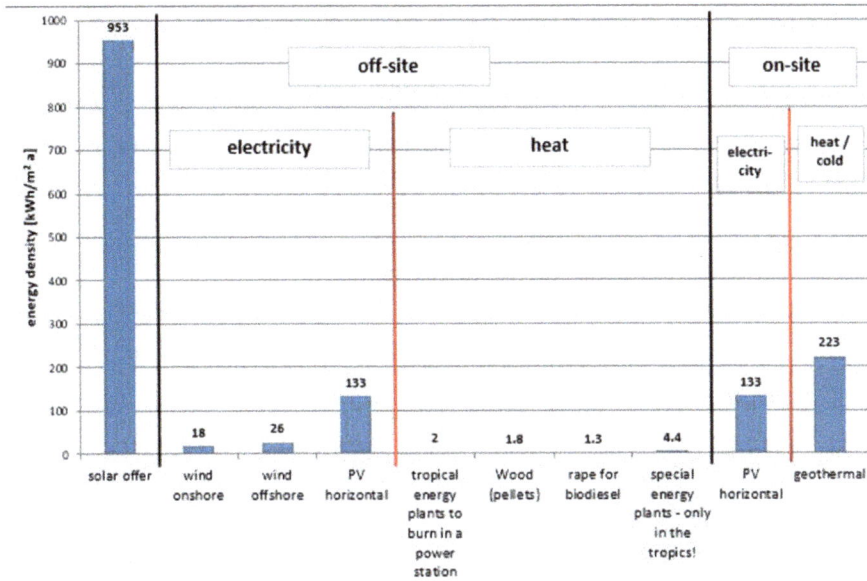

Figure 1: Energy density of different renewable energy systems. Data for solar offer and resulting harvest of PV refer to Hamburg, Germany. The value for geothermal is just to show the dimension for comparison.

borehole heat exchangers with a distance of 7 m. With the assumptions that 1 m of heat exchanger can deliver 600 Wh/d thermal energy and that the system is running 6 months a year (heating or cooling season), an energy density of 223 kWh/m² y can be derived. This value is only calculated to have a rough impression of the potential of geothermal systems in comparison to the other systems that are included in Fig. 1. Of course, the real potential of the geothermal system will be different for different locations. It is calculated on a daily basis comparing the daily heating/cooling demand and the maximal capacity of the geothermal system.

3.3 The international style room

In a first step, the standard office room was assumed as realized in common architecture, as an 'international style' room:

– N-S orientation
– Fully glazed and sealed facades with double heat protection glazing
– Internal shading system
– Air-conditioning (26°C), mechanical ventilation and artificial light during the whole time of usage

Its energy demand for all the 15 locations was simulated with Primero-Comfort [5], a transient simulation software. Results show that for all locations, more or less, it is not possible to reach a net ZEB with a satisfying urban density; if such a density is assumed, the resulting need for compensating land would be immense. That delivered conviction and motivation that the room should be optimized and adapted to the climate to reduce its energy hunger and to improve the chances to reach a net ZEB or a ZEB_CM with satisfying conditions.

3.4 Optimization of the standard office room to a climate adaptive one

To find the way to an optimized and adaptive building, several sources were used:

- Climate consultant software [6] is used to understand the temperatures, sun shading, sky cover ratio, wind velocity and humidity.
- Design strategies were derived based on the rules given by the course supervisors and those from the climate consultant.
- Vernacular architecture and best practice examples.
- Also, the opinions and suggestions of the students from these locations were considered.

Firstly, based on the climatic conditions, it was to be decided which month of a year the building can run in which of the following modes:

- Adaptive: Indoor comfort can be maintained with only natural ventilation and heating.
     It is assumed that a building can be run adaptively if the monthly mean values of outdoor temperature lie between 10 and 23°C.
     Cities like Addis Ababa, Mexico-City, Sydney and Santiago have a high potential to be run the whole year adaptive.
- Air-conditioned: Indoor comfort can be maintained only with mechanical ventilation and cooling.
     It is assumed that a building can be run only air-conditioned if the monthly mean values of the outdoor temperature is above 23°C to avoid indoor temperatures out of the comfort range.
     Singapore which has a hot and humid climate with an average yearly temperature of 27°C uses air-conditioning for 12 months of the year. Santo Domingo, Jakarta and Dar Es Salaam also need air-conditioning nearly throughout the year.
- Heating: Indoor comfort can be maintained only with heating and mechanical ventilation with heat recovery to reduce energy demand.
     It is assumed that a building can be run only in this mode if the monthly mean values of outdoor temperature lie below 10°C.
     Reykjavik, Oslo, Chicago, Beijing and partly Hamburg have such strong winter periods. But there is no location where that mode is necessary for the whole year.
- Hybrid: Indoor comfort can be maintained seasonal by running the building adaptively, with air-conditioning or heating.
     Buildings could adapt to the surroundings for a few months and depend on cooling or heating and mechanical ventilation for the rest. Chicago is the best example; it can be adaptive for 6 months and needs mechanical ventilation/heating for the rest of 6 months of the year. Hamburg and Beijing also lie in the category of hybrid.

Finally, the architecture could be adapted to the chosen modes (Fig. 2):

- Adjustment (reduction) of room depth for better natural cross ventilation (if of advantage – locations with weak wind velocities) while retaining the area of usage of 168 m² (an increase of room width).
- Reduction of the window to wall ratio in a way that overheating protection and daylight access is in good symbiosis. The ratio for locations with a dominant cloudy sky is about 50%, and for a dominant clear sky, it is 35%.

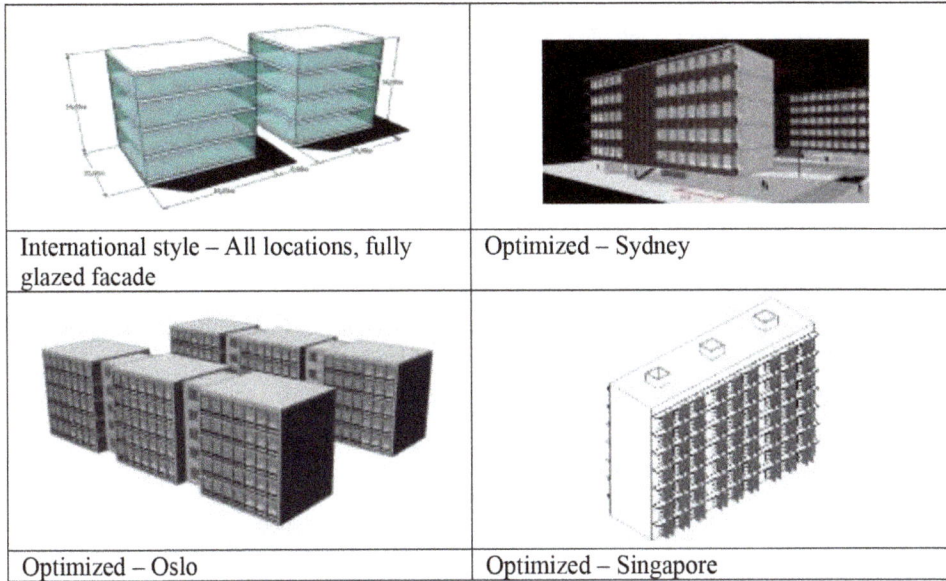

International style – All locations, fully glazed facade	Optimized – Sydney
Optimized – Oslo	Optimized – Singapore

Figure 2: Examples of international style (to find in any city) versus optimized and adaptive buildings for different locations and climates (students work).

- Convenient (external) shading system, glazing and thermal insulation.
- Intelligent size and placement of (operable) windows for daylight and natural ventilation.
- Natural ventilation strategy, especially for night cooling.
- Artificial light can be switched off if daylight is sufficient (500 lx).

These rooms were presented at the end of the university course and then further developed by the author to guarantee that they are optimized for all locations at a corresponding level. Finally, the energy demand for these rooms was simulated with the same software.

4 RESULTS

4.1 CO_2 reduction potential of optimized buildings

Figure 3 gives an overview of the possible reduction potential in primary energy (primary energy factor for heating 1.1, for electricity 3.0) between an international style building and a building that is optimized and adaptive. It can be seen that the potential (and thus the reduction of CO_2 emission!) is enormous.

It is also evident from Fig. 3 that the reduction potential is different for different locations. If cooling demand is caused by solar heat gains, it can be reduced (or brought to zero, Hamburg, Oslo, Sydney) by architectural means. If it is caused by temperatures above the comfort range, architectural means can hardly help, and cooling is necessary (Jakarta, Delhi, etc.). If the location is far from the equator, there are many hours of usage where it is dark outside – the demand for artificial light cannot be reduced (Reykjavik, Oslo, etc.).

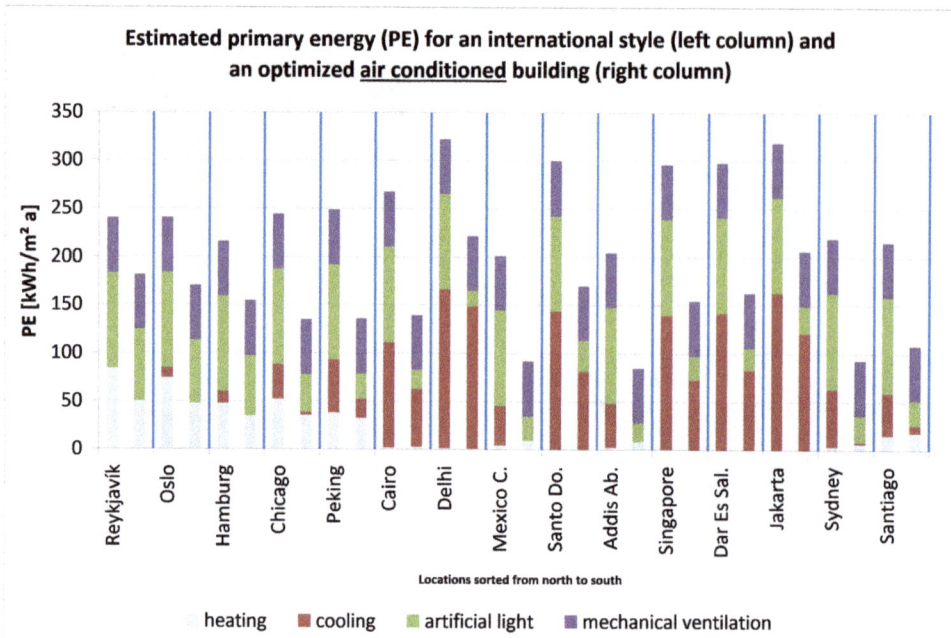

Figure 3: Comparison of the primary energy demand for international style and optimized building.

To deliver a fair comparison, it is assumed that the optimized building has mechanical ventilation and is cooled too (if necessary). Here is further optimization potential; buildings can often be run in adaptive mode (see section 3.4) and without mechanical ventilation.

4.2 Maximal urban density (plot ratio) for net ZEBs in the chosen cities

It can be calculated up to which urban density (plot ratio), net ZEB is possible for the different locations. The prerequisite that the energy demand must be covered by renewable energies gained on site sets two limits:

– The yearly harvest of a PV module system on the building's roof can cover the electricity demand (mechanical ventilation, artificial light, heat pump) for a certain number of storeys.
– A borehole heat exchanger system in the ground can deliver a certain amount of power for heating or cooling. With it, the heating and cooling demand of a certain number of storeys can be covered. For the calculation, a building distance of 20 m is assumed, which corresponds to a typical and realistic street width. From this assumption and the size of the estate, the size of the geothermal system can be determined. In case it is insufficient and there is a reserve in the PV electricity production, standard chillers (COP = 1.5) are further assumed until the PV system is exhausted.

The dominant of the two criteria decides about the maximal number of storeys. The resulting urban density can be finally calculated. Figure 4 shows the results.

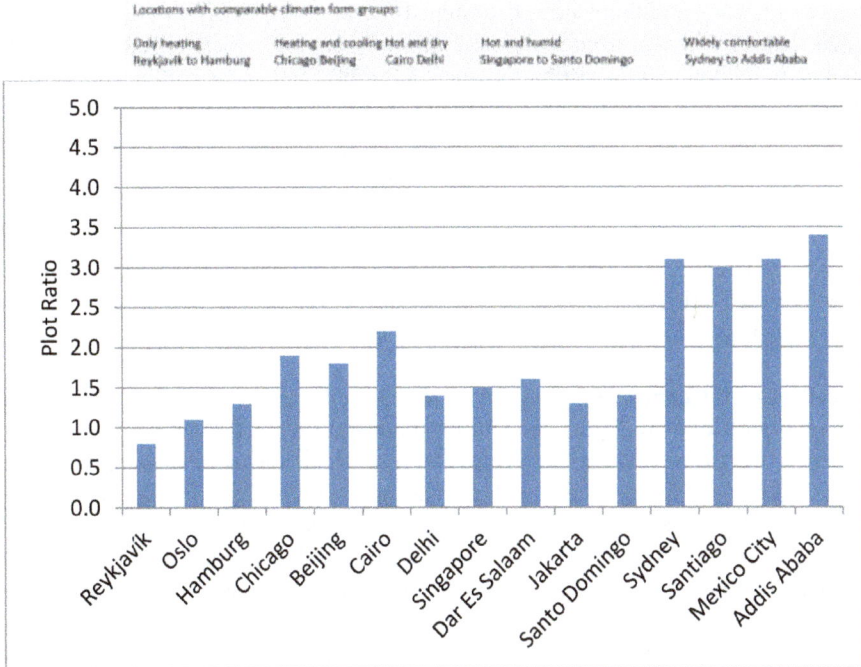

Figure 4: Maximal urban density (plot ratio) for net ZEB for the chosen cities.

From Fig. 4, it is noticeable that Reykjavik has the lowest plot ratio 0.8 which reflects unfavourable conditions prevailing in the city: a strong winter causing high heating demand, extended times of darkness causing demand for artificial light and together with the mechanical ventilation leading to high electricity demand – on the other hand, it has the lowest solar radiation and the sun position hardly above the horizon that results in a very low harvest with PV modules on the roof.

The highest possible plot ratio of 3.4 is shown by Addis Ababa which has a moderate climate. A moderate climate requires nearly no thermal energy for heating or cooling, daylight in the whole time of usage, no mechanical ventilation leading to a minimal electricity demand that can easily be covered with the PV system on the roof receiving a high amount of solar radiation. It is noted that the locations with similar climates show also similar results here. Or, vice versa, the local climatic conditions determine the chances to reach both, net ZEB and a higher urban density.

4.3 Need of compensating land for ZEB_CM for selected urban densities (plot ratio) for the chosen cities

If the planned urban density exceeds the limit for net ZEB, supplementary compensating land for renewable energy production is necessary. That land must be located outside the city (or on the sea in case of offshore wind turbines). It must be evaluated precisely if and where that is available. Competition between energy and food production must be avoided. On the other hand, a city with a high urban density avoids urban sprawl and saves land; a compact city has short distances and saves energy for transportation. The best combination for each location must be found by balancing all impacts.

The need for compensating land is determined for the urban densities (plot ratio) of two, three and four. It is known from section 4.2 that several cities reach such a high plot ratio partly already without compensating measures but a plot ratio of 4 requires compensating land for all. The results present the total land use, estate plus compensating land.

4.3.1 Compensating measures wood pellets (heat) and horizontal PV (power)
The most common systems to produce renewable energies are selected:

- For thermal energy heating, wood pellets out of renewable forestry to be transported and burned on site.
- For electricity, PV modules mounted horizontally.

Logically, for all locations, the land use for the estate decreases with increasing plot ratio. But for high plot ratios, a higher amount of compensating land is necessary, causing an increasing or decreasing total land use (Fig. 5). For the locations without significant heating demand, the total land use decreases with increasing plot ratio (Cairo to Addis Ababa).

But for other locations (Reykjavik to Beijing), the compensating land and with it the total land use increase remarkably with increasing plot ratio. The reason is that these locations have a heating demand that increases with increasing plot ratio and less of this demand is covered with the geothermal system on site. In comparison to a geothermal system, the efficiency of renewable forestry was recognized as very low (see section 3.2.3). Therefore, more energy for heating must be produced on compensating land, so higher the need of land for

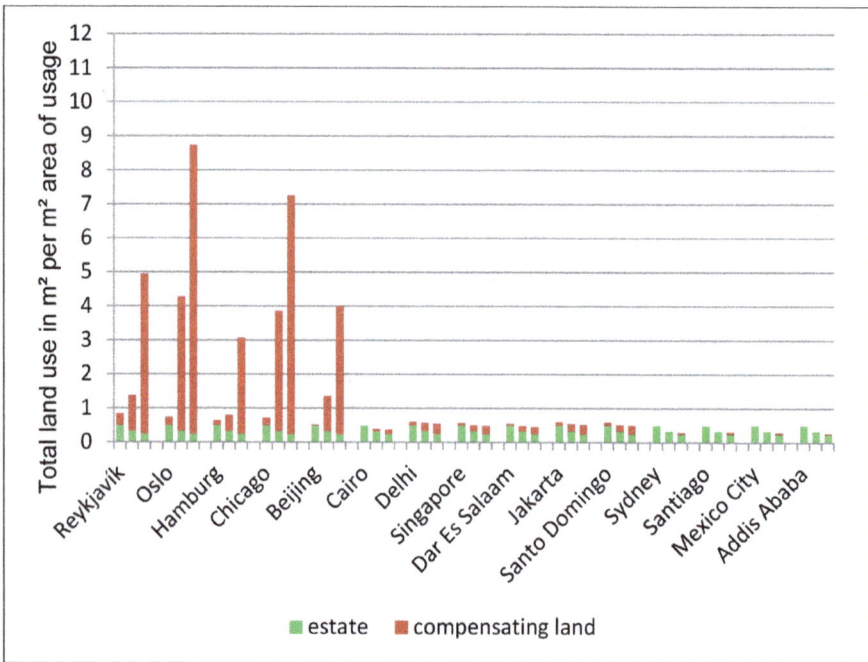

Figure 5: Total land use (estate plus compensating land) in m² per m² area of usage with compensating measures wood pellets and PV. The three columns for each city correspond to an urban density (plot ratio) of 2 (left), 3 (middle) and 4 (right).

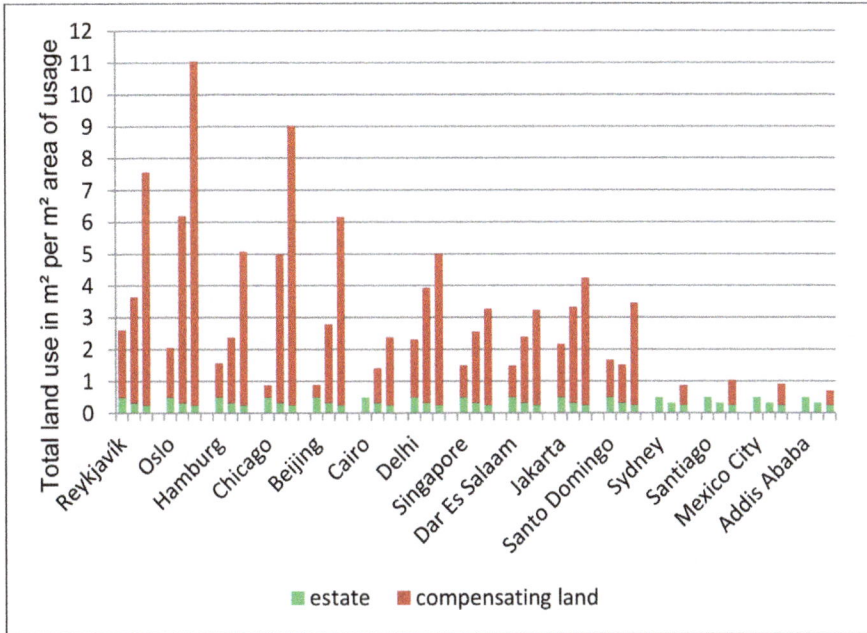

Figure 6: Total land use (estate plus compensating land) in m² per m² area of usage with compensating measures wood pellets and wind turbines onshore. The three columns for each city correspond to an urban density (plot ratio) of 2 (left), 3 (middle) and 4 (right).

renewable forestry. That effect is absolutely dominating. It must be decided for each location if such an amount of forest would be available or not.

This paper gives information to the theoretical need of compensating land. Further, it might be important if and how much compensating land would be available in practice, for certainly there are differences between Reykjavik, say, and Singapore.

4.3.2 Compensating measures wood pellets (heat) and wind onshore (power)
While there is no alternative to the production of wood pellets for heating, the impact of wind turbines onshore instead of PV modules can be investigated.

Now there is a clear effect that the total land use increases with increasing plot ratio for all chosen cities (Fig. 6). The reason is that PV modules with a higher energy density are replaced by wind turbines with a lower one. On the other hand, as already mentioned, multifunctional land use like agriculture is possible where wind turbines onshore are placed.

Only for locations with moderate climate and (nearly) no demand for heating as well as power, this increase is moderate and could be tolerated (Sydney to Addis Ababa).

5 CONCLUSIONS
A city is by definition an organism where a lot of functions are concentrated on a small space. It is not self-sufficient; energy, water, food, goods, etc. must be delivered to the city; waste, products, knowledge, culture, etc. are delivered backward to the surrounding. To find the best sustainable solution, the city and its surrounding must be regarded together. Certainly, the

city should contribute towards sustainability to unburden its harmful effects and resource depletion in the surroundings and thus the environment. Sufficiency should be the leading principle: reduction of demand by saving energy, material, water, etc. The efficiency of renewable systems to cover the remaining demand is the second principle.

Higher urban densities are of advantage: no urban sprawl, a good precondition for public transportation, bike distances.

Depending on the local climatic conditions and the chosen renewable energy systems, the land use for compensating measures can increase rapidly with increasing plot ratio, with negative impacts.

The best strategy is now as before to cover the maximum energy demand in the city. Geothermal systems have the highest energy densities for heat and cold and PV modules for electricity. Both systems should be used in all possible conditions. The best place for PV modules is the building's roofs in the city.

If compensating measures are necessary, it is important to regard the energy density of the different systems and to choose an option that does not cause extended land sprawl. But the priority is to reduce the energy demand of the buildings which means designing the buildings optimized to the climate and adaptive.

The variant international style building with renewable energy systems is not the real sustainable solution; it would lead to further exhaustion of the resources of our planet. Thus, the architecture of the 21st century is different, and it is climate adaptive.

REFERENCES

[1] Towards a zero-emissions, efficient and resilient buildings and construction sector; Global status report for building and construction, 2019. IEA and UN environment programme. https://www.worldgbc.org/sites/default/files/2019%20Global%20Status%20Report%20for%20Buildings%20and%20Construction.pdf, accessed on: 21 December 2020.

[2] Energy efficiency for buildings, UNEP. http://www.studiocollantin.eu/pdf/UNEP%20Info%20sheet%20-%20EE%20Buildings.pdf, accessed on: 21 December 2020.

[3] Towards a zero-emission, efficient, and resilient buildings and construction sector; Global status report, 2017. IEA and UN environment programme. https://www.worldgbc.org/sites/default/files/UNEP%20188_GABC_en%20%28web%29.pdf, accessed on: 21 December 2020.

[4] MacKay, D., *Sustainable energy – without the hot air*, 2008. http://www.withouthotair.com/, accessed on: 21 December 2020.

[5] HafenCity University Hamburg, Research group REAP, Dietrich, U. Primero software (version 2.0, 2018). www.primerosoftware.de, accessed on: 21 December 2020.

[6] Climate Consultant software (version 6.0, 2018). http://www.energy-design-tools.aud.ucla.edu/climate-consultant/request-climate-consultant.php, accessed on: 21 December 2020.

DESIGN AND DEVELOPMENT OF A TOOL FOR SELECTING OPERATIONS TO OBTAIN BIOMETHANE FROM BIOGAS FROM DIFFERENT SOURCES

LUCÍA GARCÍA GÓMEZ[1], SUSANA LUQUE[2], A.M. GUTIÉRREZ[3] & J.R. ARRAIBI[3]
[1] University of Oviedo, Spain.
[2] Department of Chemical Engineering, University of Oviedo, Spain.
[3] Basque Country University UPV-EHU, Spain.

ABSTRACT

Biomethane is a real alternative to natural gas and a clean way to valorize biogas obtained from organic waste in landfills or manure. It can be used as fuel in vehicles and boilers or injected in natural gas grid. Nevertheless, in Spain, biogas and biomethane are starting to be considered as an alternative to natural gas. A good way of promoting these renewable energies is supporting small and cheap treatment plants near to the place where the biogas is produced and where the biomethane can be used on-site, fostering the circular economy. An easily usable simulation tool for selecting the best sequence of unit operations for treating biogas has been designed. Modelized operations are absorption (with water, chemical and physical), pressure swing adsorption, membranes and dehydration. A step to determine if the biomethane obtained is suitable to be injected into the grid according to current Spanish regulation has been developed. To complete the design, a conventional simulation software can be used. The tool gives information about costs, consumptions and environmental impact of each selection. Pollutants modelled are those more common in biogas coming from manure although other contaminants hardly removable are considered: CO_2, CH_4, NH_3, SH_2, CO_2, O_2, N_2, H_2O and siloxanes. Unit operations have been modelled separately, and operating conditions can be easily modified by user. Some alarms have been settled to help user to make a correct selection. With this solution, a compromise between cost (compared to commercial solutions) and accuracy is met. This tool was used as a first step to design a flexible and portable prototype for treating small flows of biogas as those produced in livestock which has been later built and is on operation. This kind of solutions could help the deployment of biomethane in Spain and help installations to reduce its emissions by valorizing a residue.
Keywords: biomethane, circular economy, simulation tool.

1 INTRODUCTION

The European Union is fostering cleaner energy to make EU a global leader in renewable energy and ensure that the target of at least a 32% share of EU energy consumption coming from renewable energy sources is met by 2030 [1]. In this sense, the renewable gases such as biogas can play an important role to make the natural gas networks greener and more environmentally friendly. Spanish regulation to inject biomethane into the natural gas networks are quite restrictive, and although there was a change in our regulation to make easier the injection of gases coming from special sources (the minimum amount of CH_4 in gas to be injected was lowered from 95% to 90%), Spanish regulations are still the most hampering in Europe [2, 3]. Countries with less constricting normative have a higher penetration of non-conventional gases, such as biogas or biomethane, because treatments to meet composition of those gases with normative are not so exhaustive and therefore are economically and technically feasible.

Spain is trying to correct this trend, and in the Plan Nacional Integrado de Energía y Clima 2021–2030 (Integrated National Plan for Energy and Climate) published last March, the role of biomethane as fuel and the importance of promote its use is enhanced [4].

© 2022 WIT Press, www.witpress.com
DOI: 10.2495/EQ-V7-N1-35-47

Biomethane is nowadays growing in importance as an alternative and 'green' fuel. It is obtained using several technologies to upgrade biogas, such as membranes, water scrubbing or pressure swing adsorption (PSA). In the Smart Green Gas project, a collaborative research and development (R&D) plus innovation study supported partially by the 'Centro para el Desarrollo Tecnológico Industrial' (CDTI) that belongs to the Spanish Ministry of Economy and Competitiveness, a hybrid flexible prototype is being designed to upgrade low flows of biogas from different sources to obtain low-quality biomethane to be injected into the Spanish natural gas grid or to use it as fuel. In this study, we developed a simulation tool for a quick selection of treatments for biogas upgrading. From this tool, we obtain a first estimation of the performance of the selected treatments to determine whether the biomethane obtained meets regulations to be injected in natural gas grid or not. We also propose a novel approach to inject low-quality biomethane (not accomplishing legal limits) using the stage of grid injection at Metering & Regulating Stations (MRS) as a blending step where biomethane is mixed with natural gas, obtaining a gas which composition is accomplish Spanish. At latest, we compare results from our tool with a prototype to adjust theoretical efficiency obtained from tool.

2 OBJECTIVES

The main objective of this work is to develop a simulation tool where the unit operations based on adsorption, absorption and permeation are modelled using the physical equations that rules their performance and data available in bibliography. With this tool, a first quick selection of best unit operations for treating biogas of a given composition to reach a certain composition, high or low, can be made. From the results of the simulation tool, a pre-design of an upgrading pilot plant is obtained. In the framework of the project, a pilot plant of 1 m^3N/h has been built combining several operations according to the results of the previous stage (simulation tool). Blending in MRS is studied as a way of obtaining high-quality biomethane starting from a low-quality biomethane. Unit operations considered and modelled are PSA with zeolites, water and soda scrubbing, chemical and physical absorption, active carbon filter, drying, membranes and grid injection. At this time, the tool is almost developed (all the unit operations are modelled), and we are working in small details or improvements of the simulation tool.

Restrictions to inject non-conventional gases into the Spanish grid are summarized in Table 1. Presence of dust or particles is not allowed. Most of the operations modelled are operated at high pressure or temperature, so auxiliary stages to adjust temperature or pressure are needed. To make simpler the use of the tool, the model automatically makes calculations to adequate temperature and or pressure and the energy consumption related to it.

3 EXPERIMENTAL

Simulation tool has been developed in Excel to make easier its usage for anyone. It has been structured in sheets. The first one 'PORTADA' (Front) is the core of the tool, as is where composition of biogas is entered and where the results of the sequence of unit operations are presented. An economical and environmental analysis is also developed and shown in this sheet. The tool was designed thinking that whether several unit operations are selected, the quality of the final gas will be higher. In sheet 'PORTADA,' the sequence of unit operations is selected. It is also possible save up to five simulations to compare results graphically. Substances modelled are those that are commonly present in biogas coming from manure and

Table 1: Gas quality specifications for gas injected into Spanish gas natural system.

Gas quality specifications for gas injected into the Spanish natural gas system (0°C, 1.01325 bar) [2]				Gas quality specifications for gas from non-conventional sources injected into the Spanish natural gas system) (0°C, 1.01325 bar) [3]			
Property	**Units**	**Min.**	**Max.**	**Property**	**Units**	**Min.**	**Max.**
Wobbe index	kWh/m^3	13.403	16.058	CH_4	mole %	90	–
HHV	kWh/m^3	10.26	13.26	CO	mole %		2
Rel. density	mg/m^3	0.555	0.700	H_2	mole %	–	5
S Total	mg/m^3	–	50	NH_3	mg/m^3	–	3
H_2S + COS (as S)	mg/m^3	–	15	F/Cl	mg/m^3	–	10/1
RSH (as S)	mg/m^3	–	17	Hg	µg/m^3	–	1
O_2	mole %	–	0.01*	Siloxanes	mg/m^3	–	10
CO_2	mole %	–	2.5	BTX	mg/m^3	–	500
H_2O (dew point)	°C 70 bar	–	+2	Microog.	–		Technically pure
HC (dew point)	°C 1–70 bar	–	+5				

Table 2: Average compositions considered in this investigation.

Gas	Livestock waste (%)	Agricultural waste (%)	Sewage sludge (%)	Urban waste (%)	Landfill gas (%)
CH_4	50–80	50–80	50–80	50–70	45–60
CO_2	30–50	30–50	20–50	30–50	40–60
N_2	0–1	0–1	0–3	0–1	2–5
H_2O	Saturated	Saturated	Saturated	Saturated	Saturated
H_2	0–2	0–2	0–5	0–2	0–0.2
SH_2	0–1%	100–700 ppm	0–1	0–8	0–1
NH_3	Traces	Traces	Traces	Traces	0.1–1%
CO	0–1	0–1	0–1	0–1	0–0.2
N_2	0–1	0–1	0–3	0–1	2–5
O_2	0–1	0–1	0–1	0–1	0.1–1
Organic compounds	Traces	Traces	Traces	Traces	0.01–06*

some landfill gas: CH_4, CO_2, H_2O, NH_3, SH_2, N_2, O_2 and siloxanes. A box 'others' has been enabled to complete composition to 100% if necessary. Companies participating in the R&D project described in the introduction give the following composition for biogas.

Models of each unit operation are developed independently in different Excel sheet. Each sheet contains information relevant about the performance of the unit operation: efficiency for every substance, working pressure and temperature. Economical costs are calculated considering maintenance costs, operating costs and investment costs. Environmental asset expressed as $kgCO_2$ equivalent is made by considering CH_4 average losses for each unit operation and calculating the equivalence of energy consumption with $kgCO_2$ eq. Calculus start by the desired concentration of CO_2/SH_2 (depending on the unit operation) in the outlet of the unit operation, and depending on the grade of depuration, the amount of solvent or adsorbent is calculated. The tool calculates automatically if it is necessary to accommodate pressure or temperature, and an estimation of the effect of these adaptations is calculated through energy consumption. Lastly, a set of alarms specific for each operation is defined so that the user of the tool is advised if one of the operations selected is not appropriate for the composition of biogas and a pre-treatment is needed.

The tool has been designed to have the main information in tab 'PORTADA'. As it has been explained in this tab is also available the information about the composition of the biomethane, consumptions and alarms, whether it is necessary to adequate pressure or temperature or not. It is also indicated if there is a redundant stage (when similar unit operations have been selected and no extra depuration is achieved), and in this sheet, there is direct access (a link to go to the sheet where that unit operation is modelled) for each unit operation. If a final step of grid injection is selected, an indication of whether the biomethane can be injected into the grid or not (for high flows of biomethane) and final composition. There is also an option to indicate the quality of the biomethane to obtain: empty or 'Sin determinar' (not indicated) is the higher quality, 'Límite legal' for legal compliance and 'Bajo' (low) to upgrade the biogas to a CH_4 concentration lower than legal limit. The tool is being developed in Spanish. A caption of this tab is shown in Fig. 1.

Once the composition of the biogas has been defined and the sequence of operations selected, the tool starts calculations and gives a result, which is shown in 'PORTADA'. The selection of unit operations can be done easily, only displaying the list of defined operations. To avoid repeating unit operations, the selected operation disappears from the list of the next stage. Up to ten operations can be chosen.

By clicking 'Grabar Simulación' (save simulation), data (inputs, outputs) are copied in tab 'Resumen de resultados' (summary of results) and represented graphically $kgCO_2$ eq., CH_4 losses, maintenance and operating costs and energy consumption. See Fig. 2.

Operating conditions of unit operations and alarms can be changed easily in the corresponding Excel sheet. If an alarm is activated for one (or more) unit operation, by clicking in the name of the operation in tab 'PORTADA', the user can move to the corresponding sheet to see why the alarm is activated.

Results of each operation selected are gathered in tab 'Cálculos etapas' (calculations), where other relevant information is also compiled (consumptions, costs, ΔT, ΔP, ...), as can be seen in Fig. 3.

Global information about the upgrading process can be seen in 'PORTADA'. As calculus are made based on the elimination of almost all the CO_2 present (there is a minimum concentration established in 1%), it is possible to obtain a biogas with '0' SH_2 as the solvent is calculated to eliminate CO_2 and it is in excess in relation to SH_2.

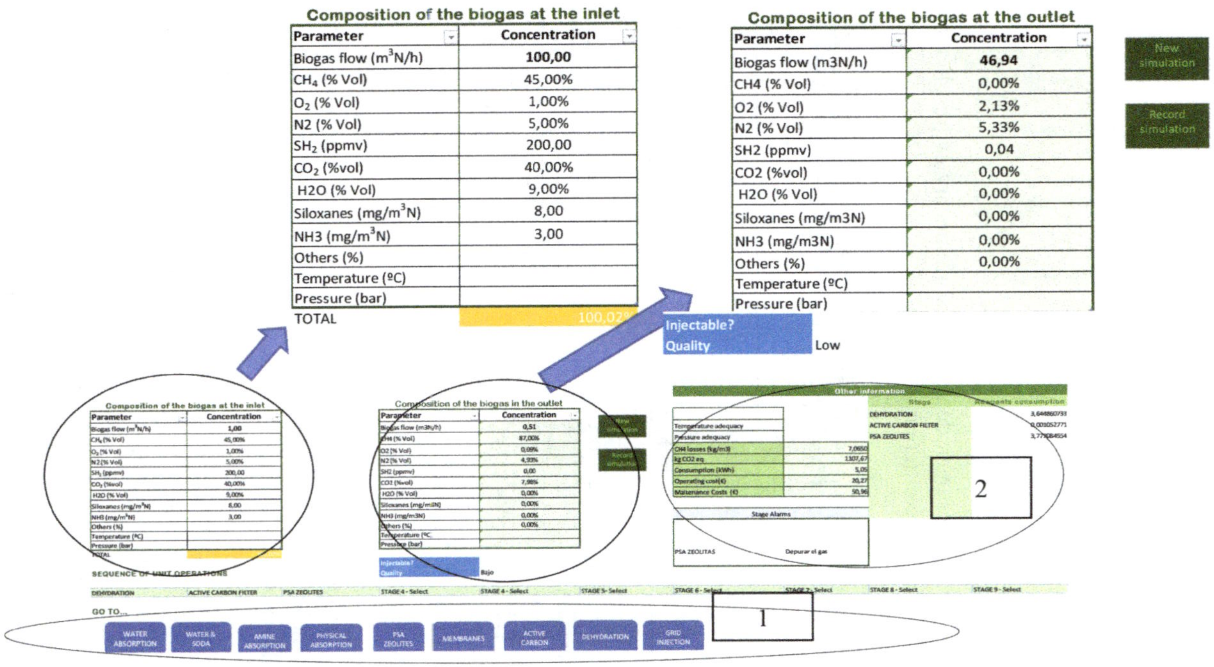

Composition of the biogas at the inlet

Parameter	Concentration
Biogas flow (m³N/h)	100,00
CH$_4$ (% Vol)	45,00%
O$_2$ (% Vol)	1,00%
N2 (% Vol)	5,00%
SH$_2$ (ppmv)	200,00
CO$_2$ (%vol)	40,00%
H2O (% Vol)	9,00%
Siloxanes (mg/m³N)	8,00
NH3 (mg/m³N)	3,00
Others (%)	
Temperature (ºC)	
Pressure (bar)	
TOTAL	100,02%

Composition of the biogas at the outlet

Parameter	Concentration
Biogas flow (m3N/h)	46,94
CH4 (% Vol)	0,00%
O2 (% Vol)	2,13%
N2 (% Vol)	5,33%
SH2 (ppmv)	0,04
CO2 (%vol)	0,00%
H2O (% Vol)	0,00%
Siloxanes (mg/m3N)	0,00%
NH3 (mg/m3N)	0,00%
Others (%)	0,00%
Temperature (ºC)	
Pressure (bar)	

Injectable? Quality — Low

Figure 1: Caption of tab 'PORTADA' (each part is amplified) with explanations in English. In this image, information related to a given combination of unit operations is shown.

SUMMARY OF RESULTS

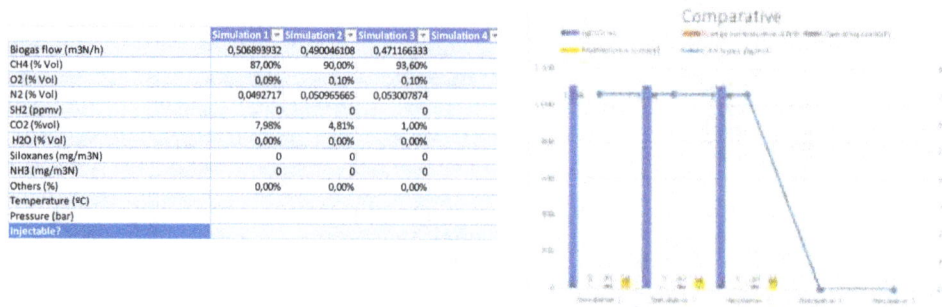

	Simulation 1	Simulation 2	Simulation 3	Simulation 4
Biogas flow (m3N/h)	0,506893932	0,490046108	0,471166333	
CH4 (% Vol)	87,00%	90,00%	93,60%	
O2 (% Vol)	0,09%	0,10%	0,10%	
N2 (% Vol)	0,0492717	0,050965665	0,053007874	
SH2 (ppmv)	0	0	0	
CO2 (%vol)	7,98%	4,81%	1,00%	
H2O (% Vol)	0,00%	0,00%	0,00%	
Siloxanes (mg/m3N)	0	0	0	
NH3 (mg/m3N)	0	0	0	
Others (%)	0,00%	0,00%	0,00%	
Temperature (ºC)				
Pressure (bar)				
Injectable?				

Comparative

Figure 2: Results shown in 'Resumen de resultados' (Summary of Results), also depicted graphically to compare different treatments.

SECUENCE	Initial 1	2	3	4	5	6	7	8	9	10	Final
Biogas flow (m3N/h)	1,00	0,920068452	0,909910819	0,506893919							0,51
CH4 (% Vol)	45,00%	48,91%	49,46%	0,87							87,00%
O2 (% Vol)	1,00%	1,09%	1,10%	0,000949774							0,09%
N2 (% Vol)	5,00%	5,43%	5,49%	0,049271701							4,93%
SH2 (ppmv)	200,00	217,3751307	3,5	0							0,00%
CO2 (%vol)	40,00%	43,48%	43,96%	0,079778525							7,98%
H2O (% Vol)	9,00%	1,07%	0,00%	0							0,00%
Siloxanes (mg/m3N)	8,00	0	0	0							0,00%
NH3 (mg/m3N)	3,00	0	0	0							0,00%
Others (%)	0,00%	0	0	0							0,00%
Temperature (ºC)	0,00	40	50	5							
Pressure (bar)	0,00	1	8	6							
ΔT (ºC)		40,00	10,00	-45,00							Temperature adequacy
ΔP (ºC)		1,00	7,00	-2,00							Pressure adequacy
m3/h solvent-kg/h reagent		3,644860733	0,001052771	3,777084554							
CH4 losses(kg/m3)		0	1,814411E-06	7,065019899							7,065021714
Kg CO2 eq		855	52,44393986	200,2303346							1.107,67
Operating consumption (kWh)		4,5	0,276020536	0,272973246							5,048993781
Operating costs		20,25	0,011040821	0,01091893							20,27
Investment costs		3000	0,003959333	2700							5.700,00
Maintenance costs		0	0,005060376	50,95500585							50,96
	1	2	3	4	5	6	7	8	9	10	FINAL

Figure 3: View of tab 'Cálculos etapas' (calculations).

For every unit operation in upper side, there is an information about the process (operating conditions, characteristics of solvent/adsorbent …). Likewise, composition of final gas and solvent, as well as the amount of reactive needed for a given concentration of CO_2 or SH_2 at the outlet gas, is shown. Compounds for which alarms have been settled as well as the limit for the alarm are also indicated there. Two buttons at the top of the Excel allow the navigation to the pages 'PORTADA' (Front) and 'Cálculos etapas' (calculations).

When calculations are made based on a physical or chemical principle, related information is also gathered. For example, in Absorption steps, Henry's Law is represented graphically to obtain a regression equation to make calculations. For adsorption, data from equilibrium are also depicted to obtain an equation to model the process. An example of this is shown in Fig. 6 for physical absorption. Some of the references used to develop the tool are [5–13], besides the specific bibliography used to model each unit operation that we don't summarize in this document because of its extension.

If there is any alarm, then the box next to 'Alarmas' (alarms) will turn on red, and in the lower side of the page, it is possible to check what compound(s) is over the limit (Fig. 6). In

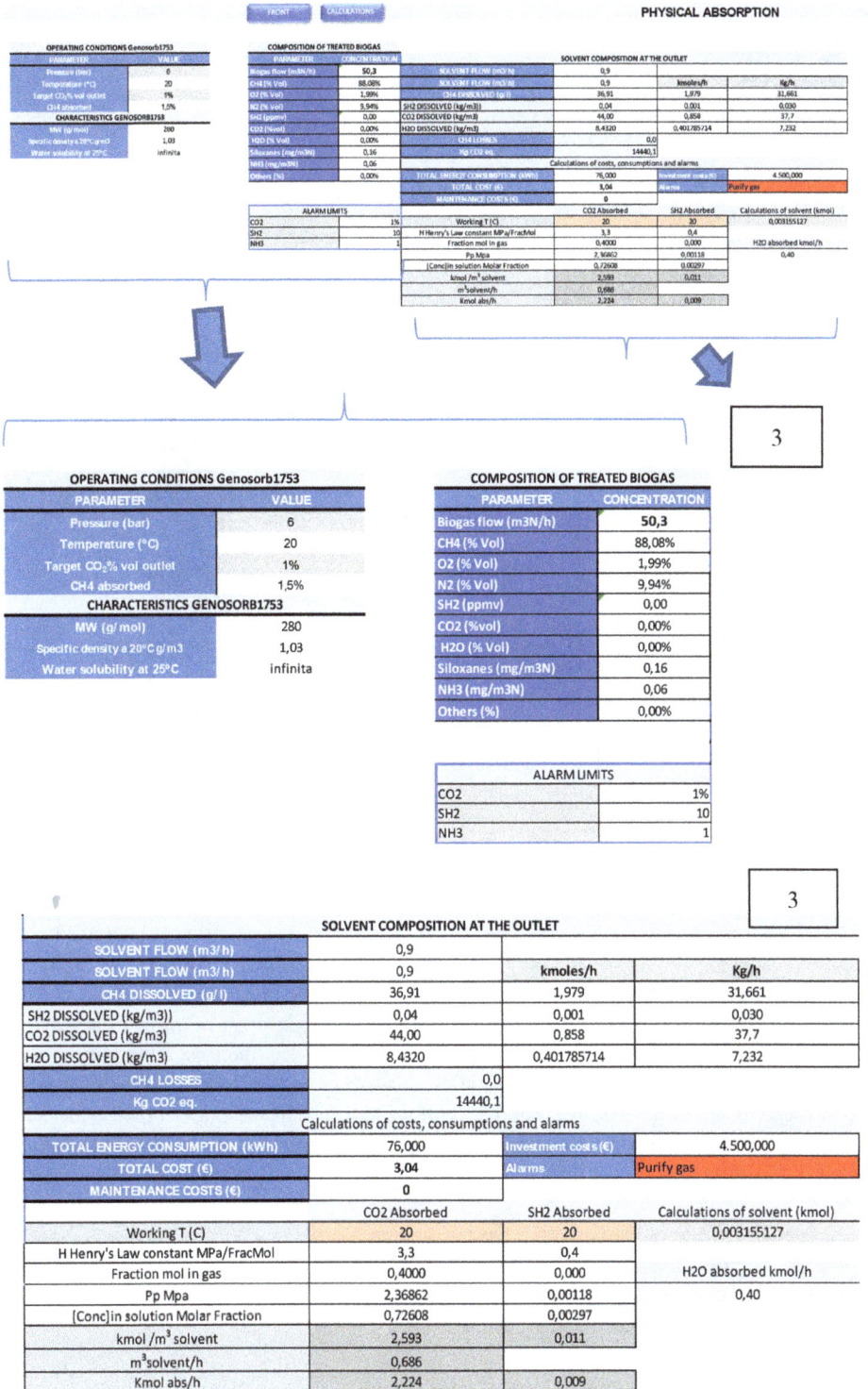

PHYSICAL ABSORPTION

OPERATING CONDITIONS Genosorb1753	
PARAMETER	VALUE
Pressure (bar)	6
Temperature (°C)	20
Target CO$_2$% vol outlet	1%
CH4 absorbed	1,5%
CHARACTERISTICS GENOSORB1753	
MW (g/ mol)	280
Specific density a 20°C g/m3	1,03
Water solubility at 25°C	infinita

COMPOSITION OF TREATED BIOGAS	
PARAMETER	CONCENTRATION
Biogas flow (m3N/h)	50,3
CH4 (% Vol)	88,08%
O2 (% Vol)	1,99%
N2 (% Vol)	9,94%
SH2 (ppmv)	0,00
CO2 (%vol)	0,00%
H2O (% Vol)	0,00%
Siloxanes (mg/m3N)	0,16
NH3 (mg/m3N)	0,06
Others (%)	0,00%

ALARM LIMITS	
CO2	1%
SH2	10
NH3	1

SOLVENT COMPOSITION AT THE OUTLET			
SOLVENT FLOW (m3/ h)	0,9		
SOLVENT FLOW (m3/ h)	0,9	kmoles/h	Kg/h
CH4 DISSOLVED (g/ l)	36,91	1,979	31,661
SH2 DISSOLVED (kg/m3)	0,04	0,001	0,030
CO2 DISSOLVED (kg/m3)	44,00	0,858	37,7
H2O DISSOLVED (kg/m3)	8,4320	0,401785714	7,232
CH4 LOSSES	0,0		
Kg CO2 eq.	14440,1		

Calculations of costs, consumptions and alarms			
TOTAL ENERGY CONSUMPTION (kWh)	76,000	Investment costs (€)	4.500,000
TOTAL COST (€)	3,04	Alarms	Purify gas
MAINTENANCE COSTS (€)	0		

	CO2 Absorbed	SH2 Absorbed	Calculations of solvent (kmol)
Working T (C)	20	20	0,003155127
H Henry's Law constant MPa/FracMol	3,3	0,4	
Fraction mol in gas	0,4000	0,000	H2O absorbed kmol/h
Pp Mpa	2,36862	0,00118	0,40
[Conc]in solution Molar Fraction	0,72608	0,00297	
kmol /m^3 solvent	2,593	0,011	
m^{3}solvent/h	0,686		
Kmol abs/h	2,224	0,009	

Figure 4: Upper side of one unit operation (physical absorption), as an example.

Henry´s Constant (MPa/ud de molar fraction) for different compounds. Xu1992   T (K) = T (C) + 273,15

T (°C) / Gas	CO₂	SH₂
25	3,57	0,44
30	3,95	0,506
40	4,67	0,641
50	5,62	0,787
60	6,55	1,01

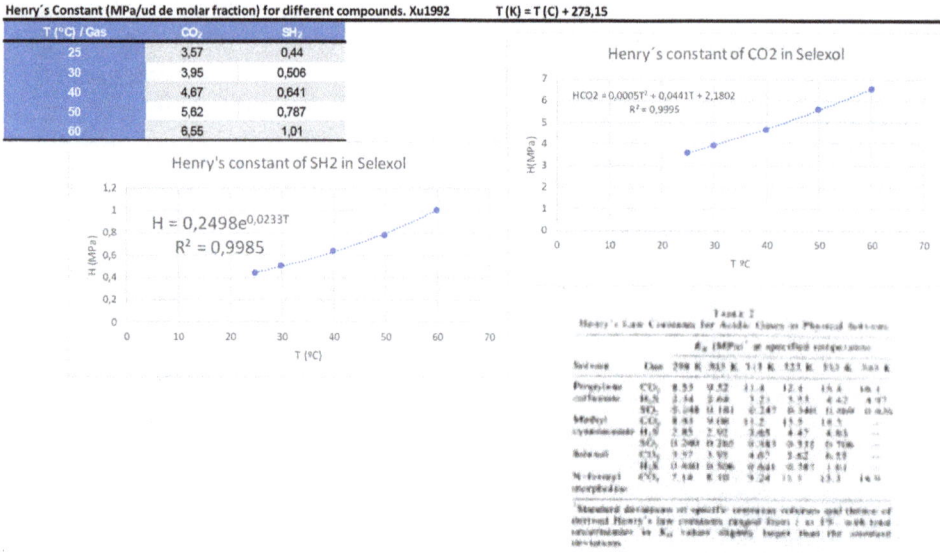

Henry´s constant of CO2 in Selexol

$HCO2 = 0,0005T^2 + 0,0441T + 2,1802$
$R^2 = 0,9995$

Henry's constant of SH2 in Selexol

$H = 0,2498e^{0,0233T}$
$R^2 = 0,9985$

Figure 5: Henry's Law for absorption of different substances in Selexol.

		m3/h solvent		0,7	0,9		Final Flow Nm3/h	
		Design factor		1,25			50,3	
Composition of biogas at the inlet		m3	kMoles	kmol absorbed	kmol outlet	m3 outlet	% Outlet
Biogas flow (m3N/h)	100,00						
CH4 (% Vol)	45,00%	45,0	2,0	0,0301	1,9788	44,3	88,08%
O2 (% Vol)	1,00%	1,0	0,0446	0	0,0446	1,0	1,99%
N2 (% Vol)	5,00%	5,0	0,2232	0	0,2232	5,0	9,94%
SH2 (ppmv)	200,00	0,020	0,0009	0,00	0,0000	0,0	0,00%
CO2 (%vol)	40,00%	40,00	1,786	1,79	0,0000	0,0	0,00%
H2O (% Vol)	9,00%	9,0	0,402	0,40	0,0000	0,0	0,00%
Siloxanes (mg/m3N)	8,00	0,00	0,000	0,00	0,0000	0,0	0,00%
NH3 (mg/m3N)	3,00	0,0	0,000	0,00	0,0000	0,0	0,00%
Others (%)	0,00%	0,000	0,000	0,00	0,000	0,0	0,00%
		Total	4,465	2,219	2,247	50,326	1,000

Figure 6: Lower part of a sheet of one unit operation.

the table, all the calculations are made including the amount of solvent/adsorbent needed and final flow of biogas.

In this project, the RMS was studied as another stage for biogas upgrading. One of the limiting aspects in Spain to inject biogas to the grid is that in most of cases to achieve the composition allowed by regulations is not affordable. To overcome this, a blending process in RMS where the flow of biomethane to inject is small in comparison to the flow of natural gas in the grid is designed to accomplish with current law in the outlet pipe of the RMS. Design to incorporate RMS as a blending system to 'purify' the biogas schematically, as shown in Fig. 7, consists of:

- A biomethane storage system with pressure regulator.
- An injection flow meter.
- A regulating electro-valve.
- A MRS in which the biomethane is injected.
- A gas sample connector.
- A wall plate with a pressure regulator, condensate filter and explosion proof barrier.
- A gas mixture dew point meter.

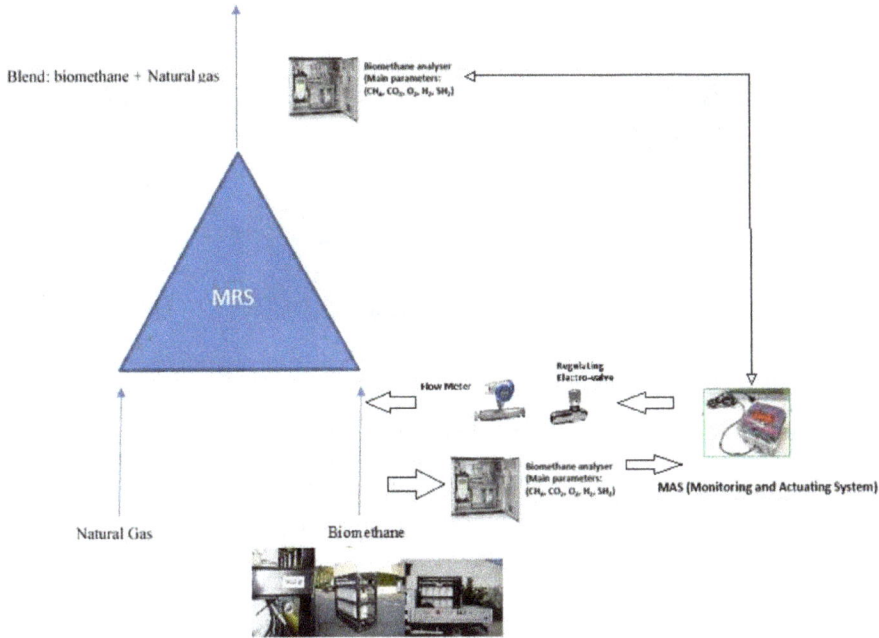

Figure 7: Scheme of the injection system.

- A gas mixture analyser of CH_4, CO_2, O_2, H_2 and SH_2.
- An intelligent continuous Monitoring and Actuating System (MAS) to inject biomethane in a smart way by controlling the gas parameters in mixture and enacting on the total volume of the biomethane injected.
- A remote server.

In order to guarantee that the mixture meets the gas quality specifications at the MRS output, we have to continuously measure the injected biomethane flow to be able to act on different parameters: if the limit value of a parameter is breached, the intelligent system will have to act on the flow rate by decreasing the amount of biomethane injected so that the mixture meets the minimum quality standards of the gas distributed in the Spanish network.

4 RESULTS AND DISCUSSION

A simulation with an average composition of a landfill gas was done. Several treatments were modelled, being the most suitable (for a cheap solution) the one combining dehydration, active carbon filter and PSA with zeolites operated at 6 bar (Fig. 8).

Composition of the biogas and the biomethane obtained from the models is shown in Table 3. Results obtained in the tool developed are shown in Fig. 9.

These results were got using the option 'Bajo' (Low) in tab 'PORTADA', to obtain a cheap biomethane that meets legal limits thanks to a final step of blending in RMS.

Results of intermediate stages are collected in Fig. 9. Thus, and according to the current Spanish regulations, the injection of biomethane into the natural gas distribution network is not feasible. See again the last column of Fig. 9. But if the gas blending solution is considered as the last upgrading stage in compliance with the Spanish natural gas regulations, then by varying the volume ratio of the mixture the permitted results are obtained in the MRS gas output. See Table 4.

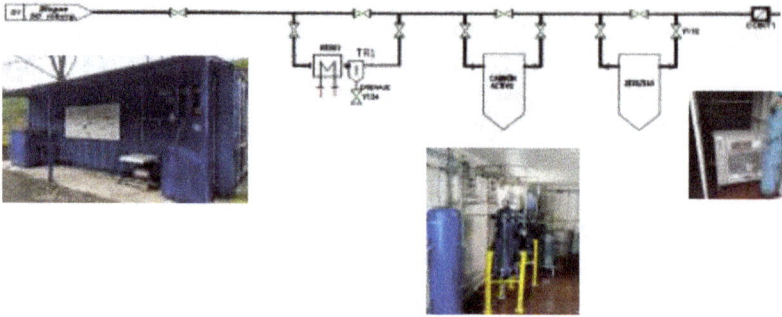

Figure 8: Prototype.

Table 3: Inlet and outlet composition of the landfill gas modelled.

Property	Units	Inlet	Outlet
CH_4	vol%	45	87.00
O_2	vol%	1	0.10
N_2	vol%	5	5.30
NH_3	mg/m³N	3	0
SH_2	ppmv	200	0
CO_2	vol%	40	7.8
H_2O	vol%	9	0

Composition of the biogas at the outlet

Parameter	Concentration
Biogas flow (m3N/h)	0,47
CH4 (% Vol)	87,00%
O2 (% Vol)	0,10%
N2 (% Vol)	5,30%
SH2 (ppmv)	0,00
CO2 (%vol)	1,00%
H2O (% Vol)	0,00%
Siloxanes (mg/m3N)	0,00
NH3 (mg/m3N)	0,00
Others (%)	0,00%
Temperature (ºC)	
Pressure (bar)	

Injectable?
Quality   Low

New simulation

Record simulation

Other information

Temperature adequacy	
Pressure adequacy	
CH4 losses (kg/m3)	7,0650
kg CO2 eq	1107,67
Consumption (kWh)	5,05
Operating cost(€)	20,27
Maitenance Costs (€)	50,96

Stage	Reagents consumption
DEHYDRATION	3,644860733
ACTIVE CARBON FILTER	0,001052771
PSA ZEOLITES	3,777084554

Stage Alarms

PSA ZEOLITES   Purify gas

Figure 9: Final results and information given by the tool.

Table 4: Composition of the natural gas + biomethane blending in the MRS gas output.

	MRS gas input	Biomethane injected in the MRS	MRS gas output
Methane (%mole)	97	87	96
Ethane (%mole)	1.1	0	0.99
Propane (%mole)	0.1	0	0.09
N_2 (%mole)	0.7	4.93	1.123
CO_2 (%mole)	1	7.98	1.698
O_2 (%mole)	0	0.09	0.009
SH_2 (ppmv)	5	0	4.5
H_2O (ppmv)	0	0	0
NH_3 (mg/m^{3}N)	0	0	0
Siloxanes (mg/m^{3}N)	0	0	0
Total (%mole)	100	100	100
Volume ratio (m^{3}N)	9	1	10
HHV (KWh/m^{3}N)	10.44	9.14	10.31
Molecular weight	16.8	18.2	16.94
Relative density	0.581	0.616	0.5845
Wobbe index (kWh/m^{3}N)	13.69	11.65	13.49
Methane number	92.7	76.6	91.09

As the biomethane composition is not in accordance with the current Spanish regulations, a last stage of blending with distributed natural gas will be used to adjust the biomethane composition to the Spanish standards on biomethane injection into the natural gas line. The quality of biogas obtained affects to treatment cost as it reduces the amount of chemicals as well as power consumption.

The prototype constructed was tested under real conditions with a landfill gas. Composition of the biomethane in the outlet of the prototype and its change with time was measured employing an Agilent 490 Micro Gas Chromatograph. We are not allowed to show data from the composition of the biogas in the inlet and the upgraded biomethane, but we can affirm that the performance of the pilot plant achieved a 95%, being the minimum molar fraction of CO_2 in the outlet next to 1% (this value is set up in the tool as final concentration to start calculations) using a zeolite 13X.

The pilot plant was designed to be portable to make it more usable and affordable, as it could be shared with nearby biogas generating facilities.

SEQUENCE	1	2	3	4	Final
Biogas flow (m3N/h)	1,00	0,920068452	0,909910819	0,471166333	0,47
CH4 (% Vol)	45,00%	48,91%	49,46%	0,87	0,00%
O2 (% Vol)	1,00%	1,09%	1,10%	0,001021793	0,10%
N2 (% Vol)	5,00%	5,43%	5,49%	0,053007874	5,30%
SH2 (ppmv)	200,00	217,3751307	3,5	0	0,00%
CO2 (%vol)	40,00%	43,48%	43,96%	0,01	94,59%
H2O (% Vol)	9,00%	1,07%	0,00%	0	0,00%
Siloxanes (mg/m3N)	8,00	0	0	0	0,00%
NH3 (mg/m3N)	3,00	0	0	0	0,00%
Others (%)	0,00%	0	0	0	0,00%
Temperature (ºC)	0,00	40	50	5	
Pressure (bar)	0,00	1	8	6	
ΔT (ºC)		40,00	10,00	-45,00	Temperature adequacy
ΔP (ºC)		1,00	7,00	-2,00	Pressure adequacy
m3/h solvent-kg/h reagent		3,644860733	0,001052771	3,777084554	
CH4 losses(kg/m3)		0	1,81441E-06	7,065019899	7,065021714
Kg CO2 eq		855	52,44393986	200,2303346	1.107,67
Operating consumption (kWh)		4,5	0,276020536	0,272973246	5,048993781
Operating costs		20,25	0,011040821	0,01091893	20,27
Investment costs		3000	0,003959333	2700	5.700,00
Maintenance costs		0	0,005060376	50,95500585	50,96
	1	2	3	4	FINAL

Figure 10: Results of intermediate stages.

5 CONCLUSIONS

Spanish regulations to inject biomethane into the natural gas grid are so restrictive that it makes the biomethane injection in most cases economically and technically unfeasible due to the high technological investment cost to upgrade biogas to the required quality levels. In this project, we propose a novel approach to overcome this drawback using the last stage of grid injection (blending) into natural gas MRS as an additional upgrading operation. We will use this stage to adapt the composition of the resulting blended gas to current strict Spanish regulations. This solution will help to the development of biomethane in Spain as a feasible renewable energy and the growth of small new projects near to the biogas source, e.g. livestock. The variability of biogas flows and quality in these cases are saved using systems for gas storage and a final step of blending with natural gas.

To this aim and as a previous stage, a simulation tool was developed to make a rapid pre-selection of the best operations to upgrade biogas to low- or high-quality biomethane given a known composition of biogas. A portable hybrid prototype was designed and constructed to test its behaviour in the upgrading of a typical landfill gas. Results of real test were like data obtained with the simulation tool that supports the theoretical models developed.

ACKNOWLEDGEMENTS

We express our gratitude to Nortegas Energía Distribución SAU and to the 'Vertedero de Lapatx' managed by Urola Erdiko Lapatx Zabortegia S.A.

REFERENCES

[1] European Parliament, «Fact Sheets on the European Union - European Parliament,» 06 July 2021. [En línea]. Available: https://www.europarl.europa.eu/factsheets/en/sheet/70/renewable-energy.

[2] Boletín Oficial del Estado, BOE, *Resolución de 21 de diciembre de 2012, de la Dirección General de Política Energética y Minas, por la que se modifica el protocolo de detalle PD-01 "Medición, Calidad y Odorización de Gas" de las normas de gestión técnica del sistema gasista*, 2013.

[3] Boletín Oficial del Estado, *Resolución de 8 de octubre de 2018, de la Dirección General de Política Energética y Minas, por la que se modifican las normas de gestión técnica del sistema NGTS-06, NGTS-07 y los protocolos de detalle PD-01 y PD-02*, 2018.

[4] Boletín Oficial del Estado, *Resolución de 25 de marzo de 2021, conjunta de la Dirección General de Política Energética y Minas y de la Oficina Española de Cambio Climático por la que se publica [...] la versión final del Plan Nacional Integrado de Energía y Clima 2021–2030*, 2021.

[5] P. A., «Biogas cleaning,» *de The Biogas Handbook Science. Production and applications*, Woodhead Publishing, 2013, pp. 329–341.

[6] P. T. H. C. T. D. Bauer F., «Biogas upgrading - technology overview, comparison and perspectives for the future,» *Biofuels, Bioproducts & Biorefining*, **7**, pp. 499–511, 2013.

[7] B. W. Beil Michael, «Biogas upgrading to biomethane,» *de The Biogas Handbook*, Woodhead Publishing Limited, 2013, p. Capítulo 15.

[8] IEA BIOENERGY, «Task 24: Energy from biological conversion of organic waste».

[9] P. N. Kadam R, «Recent advancement in biogas enrichment and its applications,» *Renewable and Sustainable Energy Reviews*, **73**, pp. 892–903, 2017.

[10] R. G. P. K. M. V. V. K. Kapoor, «Evaluation of biogas upgrading technologies and future perspectives: a review,» *Environmental Science and Pollution Research*, 2019.

[11] N. R. Kohl A., *Gas Purification*, Houston: Gulf Publishing Company, 1997.

[12] J. D. S. P. Niesner J., «Biogas Upgrading Technologies: State of Art Review in European Region,» *Chemical Engineering Transactions*, **35**, pp. 517–522, 2013.

[13] D. M. V. H. Ryckebosch E., «Techniques for transformation of biogas to biomethane,» *Biomass & Bioenergy*, **35**, pp. 1633–1645, 2011.

HOW TO FOSTER FRUITFUL COLLABORATIONS – THE IMPACT OF SUSTAINABILITY SCIENCE

LARA BRUNGS, KATHARINA KÖTTER-LANGE, JANA KOTTMEIER, REBECCA POERSCH &
PETRA SCHWEIZER-RIES
Lehr- und Forschungslabor Nachhaltige Entwicklung (LaNE, Teaching and Research Lab for
Sustainable Development), University of Applied Sciences Bochum,
Germany/Accompanying LaNE-Team Energiewendebauen.

ABSTRACT

One of the most urgent topics of the present, yet only slowly and arduously proceeding, is the energy transition, especially in the heat and building sector. Here, the basic hypothesis is that collaborations between all stakeholders involved are crucial to accelerate the process. The opening of every individual towards the perspective of others and an overarching joint intention is needed. Working from a sustainability science perspective, it is argued that approaching the transition from inside the system will lead to a common ground for collective action. The authors' role as communication researchers and transformative scientists is embedded in the broader accompanying research for the energy transition in the building sector ('Energiewendebauen'). With this paper and authors' work within the broader research network, an attempt is made to open the minds for innovative ways of working and facilitating the shift between science and practice by fostering thriving collaborations applying communication and collaboration knowledge. For this purpose, a multitude of different methods are drawn upon, some of which will be presented in this paper with a special focus on Generative Scribing, which is an artistic approach established in the context of Theory U. Although the method might initially be perceived as being rather unusual by some people and the practicing of this new way of working and communicating might even be rejected by a few, first findings show that when this method is used, people are intuitively attracted and open up in the process.

Keywords: collaboration, energy transition, generative scribing, interdisciplinarity, sustainability science, Theory U, transdisciplinary, transformation, transformational science, transepistemic.

1 INTRODUCTION

Smart grids, fuel cells in electrical power systems, energy-efficiency retrofits – thanks to continuous research we do already have the solutions at hand, yet we struggle implementing them. One example demonstrating the tedious development of the energy transition related to the authors' specific project is Germany's refurbishment rate stagnating at 1% making it highly unlikely to reach the federal government's target of 'an almost climate-neutral building stock' until 2050 [1, 2]. The energy transition plays a vital role in achieving the goals of the 2030 Agenda for Sustainable Development [3] and is a fundamental part of the 'great transformation' called for by the German Advisory Council on Global Change (WBGU) [4].

United Nations Development Programme (UNDP) [5] and many others [6, 7]. Even though positive developments have been realized since the adoption of the Sustainable Development Goals, the effort put into attaining them is still not sufficient as 'the world is not on track for achieving most of the 169 targets […]' [8]. The transition of the energy system towards sustainability is considered as a 'wicked problem' [9], meaning that it is highly complex due to the numerous actions involved and the interconnectedness between them. This makes it nearly impossible to grasp in full detail or even to oversee it. Hence, we are convinced that there is an urgent need for transepistemic and transdisciplinary collaborations between science, practice, politics, and the society as a whole to stimulate the development and to tackle the challenges ahead. To establish the collaborations needed, it is suggested that inclusive, appropriately uncommon methods like Art of Hosting [10], Design Thinking [11], or Generative Scribing [12] can be used in order to engage ourselves in a 'generative dialog' [13] resulting in collective action and shared responsibility with the aim of mitigating the climate crisis.

2 TRANSITION RESEARCH AND COMPLEX SYSTEMS APPROACHES

A need for change towards sustainability on the local, regional, national, and global level, i.e. transformation/transition, has been voiced in varying contexts. In this paper, transformation and transition are used interchangeably. As Göpel, like others, has put it: '… when it comes to defining what constitutes a transition versus what constitutes transformation (…), there is not much difference' [14]. The field of study that focuses on this topic is referred to as transition research. The origins of this field can be traced back to the 1990s, and it has grown into '[…] a highly multi-, inter- and transdisciplinary field […]' [15]. During the years, its development has been fuelled due to the rising awareness that new approaches are needed, so that complex societal challenges and their dynamics can be investigated and adressed by guiding the development of social systems [15]. At first, mainly socio-technical systems and the transitions happening in these have been analyzed, however this focus has been extended by incorporating societal systems in general [15]. The starting point of transition research is a 'persistent or wicked problem', of which the identification and understanding of key drivers and stabilizers within the system is aspired to [14]. In institutionalism theory, different stabilizers are examined as 'path dependencies' [14]. These are important for identifying processes in systems that impede change. From a complex systems perspective, it can be said that they inherit essential feedback loops [14].

There are many different approaches in transition research. One of the most popular frameworks within the socio-technical approach is the so-called multi-level perspective (MLP) [15]. It has been put forward by scientists aiming at synthesizing socio-technical systems and evolutionary economics [16]. According to Geels, it is a '[…] mid-range theory that conceptualizes overall dynamic patterns in socio-technical transitions' [17]. Within this framework, transitions are described as non-linear processes, which are the result of interactions between different levels and are defined as '[…] a shift from one regime to another regime' [17]. The MLP differentiates three levels, all of which consist of heterogenous elements called 'technological niche', 'socio-technical landscape', and 'socio-technical regime' [16]. The last one is the most important as the others are defined in relation to it [17]. It is made up of rules that serve as coordination and orientation tools for a social group such as shared beliefs, lifestyles, user practices, or legally binding contracts [17]. The first level is a place, in which small groups work on innovations that are different to dimensions of the regime and actors hope for the innovations to diffuse to regime level [17]. However, this is not an easy task, but once networks become larger and a 'dominant design' is the result of aligning learning processes, niches gain impetus [17]. The second level describes the broader context, which does not only include the technical and material background but also political ideologies, societal values, macro-economic patterns, etc. [17]. The MLP abolishes simple causalities and emphasizes 'circular causality' due to the absence of only one driver leading to transitions [17]. This underlines the systemic root in the second order cypernetics of this predominantly technological approach.

Göpel has put forward her own approach based on the MLP and the multi-phase concept, in which she emphasizes the importance of humans in transformation processes, as they are '[…] both subject and object of making history, how reality today shapes the imaginary of how reality could be in the future' [15]. She highlights that how people behave in the world is influenced by how they see it [14]. Meadows is of the same opinion and argues that 'paradigms are the sources of systems' [18]. Göpel however refers to 'paradigm' as 'mind' and argues that it stresses the process of knowing rather than the idea itself. Mind expresses that 'the way that seeing and believing differently goes beyond an update of information. It also means changes in attention, consciousness, instinct, imagination, judgment, power, sense, spirit, and psyche' [14]. Göpel extents the MLP by linking it with this reflexive ontology and the multi-phase concept, which is helpful when looking for the best stage to implement deliberate change in complex systems [14]. She calls this concept 'radical incremental transformation' [14] and stresses that 'the source of intentional change is human thinking, feeling and acting' [14].

Our work is based on complex system's approaches, which try to advance the challenges arising due to the high complexity of 'wicked problems' like the energy transition in a holistic way, taking into account technological, economic, political, socio-cultural, and, not least, ecological dimensions, as well as the interactions between them [19, 20]. Systems are described as 'composed of multiple components of different types, both tangible and intangible. They include, for example, people, resources and services, as well as relationships, values, and perceptions' [21]. The system approach is also often explained by modern quantum theory. After Werner Heisenberg, this states that 'matter and energy [...] form in the end an inseparable whole. Not the material particles are decisive in the microcosm, but the effective fields between them' [22]. In other words, within a system (be it the energy sector, our society, or the whole world) equally important to the single components are the relationships between them [23]. The term 'emergence' describes the occurrence of properties of a system 'that are not given with [its] objects but result from their interactions' [24]. Therefore, it is important to consider the process as a whole, considering the individual as well as the relations between them.

3 THE ENERGY TRANSITION AS A TRANSEPISTEMIC AND COLLABORATIVE TASK

At first sight one might think that the energy transition is mainly dependent on technological innovations. However, as shown above, not only technological feasibility is relevant but also a variety of different factors from financial to socio-cultural ones and the most important aspect: the human being. People accepting and supporting the process are needed to realize the transformation successfully, sustainably, and healthily.

In order to break existing path-dependencies, equally indispensable in all factors is – more than the dissemination of knowledge – the collective will of people in different sections of society to collectively achieve our aim. The whole process of the energy transition is far too complex and multi-dimensional to be solved by singular disciplines. Each discipline may add one dimension to the view on the transition, but to get a full picture of what needs to be done, several perspectives are required including those of stakeholders outside science. It is here argued that it is important to overcome silo mentality and instead connect actors and expertise from the different parts of the 'system' energy transition. A sole focus on disciplinary thinking in fact omits truly innovative, creative, and – above all – holistic ideas by narrowing perspectives on crucial topics. If the energy transition is to be realized, recipients of new energy systems and all relevant societal members willing to participate must be recognized as fully legitimated to contribute to this process [25]. Experts not only build differing know-how, but they also often experience varying layers of our society's shared reality and, as to our human nature, experience them differently. As everybody carries his or her own set of experiences of values and interests, it is a complex interplay that shapes our individual perceptions and ways of thinking and, evidently, also how we approach challenges and find solutions. For this reason, we pledge not only for inter- and transdisciplinary, but also for transepistemic collaborations [26]: including various ways of knowing into the joint effort to transform not only but also our energy system, e.g. practical knowing as well as various types of scientific knowledge.

These collaborations do not only allow the transfer of existing knowledge from science to practice, but they also facilitate the mutual generation of new knowledge. Real collaborations come about when people connect on a deep level working towards a mutual goal on which everybody can agree. To establish such collaborations, 'safe' spaces with an open communication need to be created, in which all participants are willing to be vulnerable and are, therefore, able to empathize with others [10]. Within this context of true collaborations, we are not led by the past, but are sensing the emerging future guiding us towards a shared intention [13].

4 SUSTAINABILITY SCIENCE SUPPORTING TRANSEPISTEMIC COLLABORATIONS

In this research within the larger project of the energy transition for the German building sector, the aim is to contribute with communication and collaboration knowledge from the sustainability science. Questions targeted are how can we create a safe space where people from different disciplinary cultures and also outside science can come together and open up to each other's realities and experiences? What might be the best approach to pave the way for socio-technical innovations in the form of societal and political acceptance, a window of opportunity, to be filled with new ideas? Which concepts are most suited to design our working spaces and communication processes for the development of connection and a shared intention? And how can we pool our expertise and efforts into what we call 'concerted action' and co-create a sustainability transformation?

Sustainability science, by nature, incorporates all research fields and dissolves disciplinary boundaries, not only between scientific disciplines but also between science, practice, politics, and the society as a whole. Sustainability science is transepistemic and transdisciplinary, value-oriented and normative, taking over responsibility and systemic [26, 27]. In addition, it can be seen as intendedly transformative [28] as well as consciousness based [27]. This kind of sustainability science is not only occupied with the analysis and description of transformation processes (transformation research) but also normatively fosters change itself (transformative science). This action research is accompanied by reflecting and learning processes [29] and deepening the consciousness not only of the subjects of research but also of the researchers themselves [27].

We as transformative and awareness-based sustainability scientists advocate for focusing on the deeper dimensions of individuals exploring the source of our actions [27] – the 'blind spot' as Scharmer calls it [13]. Scharmer and Käufer call this as 'awareness-based action research' [30]. In this kind of research, science does not only do research, but it also seeks to change with the society in the process [27].

5 OUR ASSIGNMENT WITHIN THE PROJECT

Our research team has been assigned with the task to promote transdisciplinary and transepistemic collaborations between the many different actors involved in the energy transition of the building sector.

Figure 1: The burning lens of the energy transition – Source: Jana Kottmeier.

Figure 1 stems from the method Generative Scribing, the application of which will be described in Section 6.2. The artwork visualizes how we imagine our work: figuratively speaking we seek to be a burning glass that unites the different ideas, technologies, and disciplines coming together as beams, which pools them into 'concerted action'. What we thrive for is to create 'Communities of Practice' [31] among people working to bring about the energy turnaround in the heat sector. The diversity of stories and perspectives represents the complexity of life, which again lies at the core of sustainability concerns.

Through organizing and planning events like workshops, professional peer-coaching or conferences, we open up possibilities and 'safe spaces', in which the participants are able to communicate and collaborate in different group sizes and settings. Systematically and participatively, we enable new collaborations to emerge and growing collaborations to come to full bloom. We seek to create opportunities for people to meet and move something together. Communication is – metaphorically speaking – the mortar between all the bricks forming the energy transition and our project team intends to help and foster a good and open communication, so that real collaborations can be established.

6 METHODOLOGY APPLIED IN OUR RESEARCH

In this section, we present a general overview of a number of concepts and methods that are generally used by organizations (like, e.g. Collaborative Helvetica) and specifically by our research team to foster these transformation processes by collaborative work structures, refer to Table 1. Moreover, one specific method called Generative Scribing will be outlined in detail. Making use of the methods requires stepping out of our personal and disciplinary comfort zone, which is naturally accompanied by resistance at first. That is why a significant amount of courage and time needs to be poured into networking, communicating, and establishing a basis of understanding and mutual acceptance.

6.1 A brief overview on methods

A lot of our methodology can be classified as what Scharmer and Käufer refer to as methods used in 'awareness-based action research' [30]. Scharmer with his diverse teams as well as research collaborators has developed a concept called Theory U that can be understood as a

Table 1: Toolbox for collaborative communication.

Approaches	Methods – how the approaches are brought into practice	Sense and purpose – what we want to achieve
Transformative Science	**Real-World Labs** Spaces of transdisciplinary as well as participatory and transformative research [33], that perform interventions in real-world social contexts in order 'to achieve a deeper understanding and the realization of transformations' [28].	**Rather a paradigm shift than an approach:** Moving from a describing science to one that initiates and catalyzes transformational change [29, 34].
Collaboratories based on the Theory U	**Generative Listening** Opening the will so that one's perspective merges with the systems' perspectives in a way that it is leading from the emerging system's future. 'When that deeper generative field is activated, we usually experience it as time slowing down, space opening, widening, the sense of self decentering, while the self- other boundary opens up to a collective presence from which the conversation seems to flow' [35].	**Enabling generative social fields** by hosting integrative conversations between the research network and the field actors. Enabling a sense of 'us' that enables the participants to co-create, as it arises through shared experience and a collective will [35, 37, 38].
	Guided Journaling A set of questions that provide stimuli for sensing and reflecting about oneself and the system around. Each member of a group goes through the predefined questions in its personal space followed by an exchange in small groups or the plenum [35].	Finding the deeper sources of what we do and creating intentions for joint action.
	Prototyping Rapid, co-creative, iterative development of prototypes that transform the future of the system into reality. Alternates with silent phases of reflection (e.g. journaling) [36]. 'As we *enact* prototypes we explore the future by doing. The relationship between observer and observed continues its inversion' [13].	Bringing the predesigned actions into reality, experimenting together what works.

Art of Hosting [10] [39]	**World Café** Table groups of four to five people are formed around a common topic. One person is the host who will be in charge of the table's outcome for the following three rounds. The topic is now discussed between the visitors. At the end of the first round, all table guests except for the host look for new tables and thus carry on the dialogs and results. In this way, the subsequent rounds, which follow the same patterns, are enriched.	Since misunderstandings are the rule and comprehension the exception [40], it is necessary to create fertile and appreciating spaces in which good communication 'happens like on its own'. Art of Hosting is more than a set of methods, it is a mindset, a vehicle for collaboration and the attempt to open up a space for emergence. The World Café is used for deeper reflection on topics we want to move further.
	Open Space Open Space is defined by the self-organization and freedom and spontaneity of the participants. Topics can be freely introduced by individuals and supervised in the course of the method. Without an own topic, participants move freely. At the end, a protocol for all topics is provided.	Open Space is meant to bring ideas together and lead them to joint action.
	Appreciative Inquiry Focusing on what works rather than on what does not work – dreams, desires, strengths, passion, and sources. Inclusion of the methodological needs and experience of the participants.	Appreciative Inquiry helps to focus on what is already there instead of what is missing.
Design Thinking	Depending on the approach, the design thinking process is described as having from three (inspiration, ideation, implementation) [11] to seven steps (define problem, understand problem, define persona, ideate, design prototypes, test prototypes, integrate prototypes) [41], with constant iterations. Divergent and convergent thinking are alternately addressed in the process. Furthermore, each step comes with its own specific methods.	Integrating users' stories, perspectives and needs into the (scientific) process. Increasing users' acceptance and participation is targeted [42].

framework, a method, and a mindset at the same time [13]. Basically, the form of the U, also serving as the eponym for the theory, visualizes the dynamic road that every single successful group process will ultimately follow (a visualization of this can be found in ref. [32]). It starts with becoming aware of one's own patterns (or those of one's organization, team, etc., as it is originally a management theory), opening them up and letting them go (the left side of the U), until at the bottom the system (person, organization, etc.) gets to connect with itself/ themselves. This is the point of connecting, not only with one's own motivation and goals but also with those of other team members and parts of the system. It is where we create the shared sources, motivations, and support in our work within the system of people engaged with the energy transition of the building sector. The right, upcoming part of the U finally is the building of the future, by prototyping new social practices, lifestyles, and sustainable governance structures. Via prototyping, which is part of design thinking inside Theory U, the emerging future can be explored and created together in action. This process enables groups to sense and create the emerging future together and leads to a shift from an ego-system perspective to an eco-system perspective [25]. If such a stage is reached, it is not important who did what, but it leads to the establishment of a mutual feeling of accomplishment – a sense of community and a collective transformation.

6.2 Helping the emerging future come to live – The application of Generative Scribing

The method Generative Scribing stems from the so-called scribing, which is a visual practice created in the 1970s in California and has been established by Bird in the context of Theory U [12]. When practicing scribing during a conference, a conversation, or a meeting, the artist or 'scribe' links topics with the artwork and, e.g. thereby supports decision making [12]. Here primarily the content on the surface is captured by the artwork. This is taken a step further by Generative Scribing as the scope of the scribe is expanded [12]. It is a form of art, a process, a social act 'with which we open to the unknown to bring it to life – of, and for, a social body' [12]. The artwork is the result of a participatory process, in which the scribe 'serves as artistic aid(s) in shared seeing and human navigation' [12]. The scribe is able to voice the different views of the social body within the artwork by sensing non-judgmentally and generatively in the present social field [12]. When applying Generative Scribing, the artist listens on a deep level and makes the essence of what is being said visible [12]. Bird has identified four levels of scribing [12] in relation to the four levels or fields of listening described by Scharmer [13]. For Generative Scribing, it is important to sense with the heart and to work in connection with the source, the inner point guiding our actions [12]. The artwork is a support for the social body to fearlessly tack and, therefore, make changes happen. It has high potential to becoming the basis of a shared intention guiding the social body through challenging times [12].

Typically, the scribe and the artwork of the scribe can be seen by the audience during the process and is, thereby, immediately influencing and transforming the thinking in the room [12]. However, in some settings, it is not possible or useful to do so, e.g. during a conference with differing settings. Bird stresses in her book that Generative Scribing is in fact changing with the changes in the social field and, therefore, the application of the method can vary [12].

The method has been applied in several different meetings since our project has started. Our experiences show that, at first, people tend to be sceptic when the method is applied. However, the artwork that developed during the meeting has always been shown at the end to the attendees to be open for reflection and discussion. Thoughts have been shared, upcoming questions were uttered, and, in further consequence, it served as an opportunity to jointly review in retrospect the essence of the meeting. We therefore argue that the application of such methods in a technologically driven social field can softly open the minds for new ways of working towards more collaborative structures as it supports the process of finding a joint intention. Results can be found in ref. [43].

7 CONCLUSION

The scientist and Lord Mayor of the 'transformation city' Wuppertal, Uwe Schneidewind, recently created a term for the ability to deal with the complexity of dimensions and the diversity of actors involved in transformation processes: for him this is the *Zukunftskunst* (roughly translated: the 'art of futurism') [44]. This notion includes some basic principles of our work concerning transformation processes: it needs dedication, space, and moreover time [16]. As a matter of course, we have experienced some setbacks, as naturally most of the resistance towards uncommon methodology (which ours is for most of our system peers), occurs in the beginning [45]. But we have also experienced the impacts of the methods applied when people take their time to accommodate themselves to something new and uncommon. In our own research team, we discovered the effect of truly connecting with each other leading to a feeling of community and power. Altogether, we are deepening our 'sense of the field' [25] and whilst working together, we continuously experience the U process.

ACKNOWLEDGMENTS

We are very thankful for the inclusion, support, and funding of our project work by the German Federal Ministry for Economic Affairs and Energy (project number: 03EWB001B), and we would like to thank the Projektträger Jülich (project carrier: Jülich) for the great support and guidance. Only the authors are responsible for the content of this paper.

Furthermore, we would like to express our deepest gratitude towards our colleagues in the broader accompanying research network for the energy transition in the German building sector and all grantees for the courage and time they are investing into exploring unknown, uncommon waters. And last but not least, we are thankful to the international U.lab community whom we are part of and feel encouraged by ourselves to support the great transformation urgently needed.

REFERENCES

[1] German Energy Agency (dena), *Building report – Energy efficiency in the building stock – statistics and analyses*, 2018, available at https://www.dena.de/fileadmin/dena/ Dokumente/Pdf/9268_dena_concise_2018_building_report.pdf (accessed 08 August 2021).

[2] Energiewendebauen. Forschung für energieoptimierte Gebäude und Quartiere, available at https://www.energiewendebauen.de/startseite (accessed 09 August 2021).

[3] United Nations (UN), *Transforming our world: The 2030 Agenda for Sustainable Development*, 2015, available at https://sdgs.un.org/sites/default/files/publications/21252030%20Agenda%20for%20Sustainable%20Development%20web.pdf (accessed 05 August 2021).

[4] German Advisory Council on Global Change (WBGU), *World in Transition – A Social Contract for Sustainability*, 2011, available at https://www.wbgu.de/en/publications/publication/world-in-transition-a-social-contract-for-sustainability (accessed 05 August 2021).

[5] United Nations Development Programme (UNDP). *The great transformation, Working with radical uncertainty in a planetary crisis, Web Site*, available at https://stories.undp.org/the-great-transformation?utm_source=web&utm_medium=homepage&utm_campaign=greattransformation (accessed 10 August 2021).

[6] Korten, D.C., *The Great Turning: From Empire to Earth Community*, Berrett-Koehler: San Francisco, 2007.

[7] Polanyi, K., *The Great Transformation: The Political and Economic Origins of Our Time*, Beacon Press: Boston, MA, 1957.

[8] United Nations (UN), *Independent Group of Scientists Appointed by the Secretary-General, Global Sustainable Development Report 2019: The Future Is Now – Science for Achieving Sustainable Development*, 2019, available at https://reliefweb.int/sites/reliefweb.int/files/resources/24797GSDR_report_2019.pdf (accessed 09 August 2021).

[9] Seager, T., Selinger, E. & Wiek, A., Sustainable engineering science for resolving wicked problems. *Journal of Agricultural and Environmental Ethics*, **25**, pp. 467–484, 2012. https://doi.org/10.1007/s10806-011-9342-2

[10] AoH Team Hamburg. *The Art of Hosting & Harvesting conversations that matter*, Online, 2019, available at https://www.einfachgutelehre.uni-kiel.de/wp-content/uploads/2016/04/Aoh-Hamburg-Handbuch-Final.pdf (accessed 08 August 2021).

[11] Brown, T., Design thinking. *Harvard Business Review*, **86(6)**, pp. 84–92, 2008.

[12] Bird, K., *Generative Scribing. A Social Art of the 21st Century.* Presencing Institute: Cambridge, 2018.

[13] Scharmer, O.C., *Theory U: Leading from the Future as It Emerges – The Social Technology of Presencing.* Society for Organizational Learning: Cambridge, MA, 2007.

[14] Göpel, M., The great mindshift, how a new economic paradigm and sustainability transformations go hand in hand. *The Anthropocene: Politik–Economics–Society–Science*, vol. 2, ed. H.G. Brauch, Switzerland: Springer Nature, 2016.

[15] Loorbach, D., Frantzeskaki, N. & Avelino, F., Sustainability transitions research: transforming science and practice for societal change. *Annual Review of Environment and Resources*, **42**, pp. 599–626, 2017. https://doi.org/10.1146/annurev-environ-102014-021340

[16] Grin, J., Rotmans, J. & Schot, J., Transitions to sustainable development: new directions in the study of long-term transformative change. *Routledge Studies in Sustainability Transitions*, Routledge: New York, 2010.

[17] Geels, F.W., The multi-level perspective on sustainability transitions: responses to seven criticisms. *Elsevier Environmental Innovation and Societal Transitions*, **1**, pp. 24–40, 2011. https://doi.org/10.1016/j.eist.2011.02.002

[18] Meadows, D., *Leverage Points: Places to Intervene in a System.* The Sustainability Institute: Hartland, VT, 1999.

[19] Geels, F., McMeekin, A., Mylan, J. & Southerton, D., A critical appraisal of sustainable consumption and production research: the reformist, revolutionary and reconfiguration positions. *Global Environmental Change*, **34**, pp. 1–12, 2015. https://doi.org/10.1016/j.gloenvcha.2015.04.013

[20] Göpel, M., Hermelingmeier, V., Kehl, K., Then, V., Vallentin, D. & Wehnert, T., *System innovation lab: shaping Europe's energy future*, 2016, available at https://epub.wupperinst.org/frontdoor/deliver/index/docId/6538/file/6538_System_Innovation_Lab.pdf (accessed 14 May 2021).

[21] Abercrombie, R., Harris, E. & Wharton, R., *New Philanthropy Capital (NPC), Systems Change, A Guide to What It Is and How to Do It*. Lankelly Chase: UK, 2015.

[22] Kurt, H., *Wachsen! Über das Geistige in der Nachhaltigkeit*. Verlag Bild-Kunst: Bonn, 2010.

[23] Harding, S., *Animate Earth. Science, Intuition and Gaia*. Green Books: Cambridge, 2009.

[24] Merkel, W., Brückner, J., Wagener, H.J. & Teil II, Theoretische Paradigmen, System. *Handbuch Transformationsforschung*, eds. R. Kollmorgen, W. Merkel & H.J. Wagener, Springer: Wiesbaden, 2015.

[25] Nanz, P., *Handbuch Bürgerbeteiligung. Verfahren und Akteure, Chancen und Grenzen*. Bundeszentrale für politische, Bildung: Bonn, 2012.

[26] Schweizer-Ries, P. & Perkins, D.D., Sustainability science: transdisciplinarity, transepistemology, and action research. *Umweltpsychologie*, **16(1)**, pp. 6–10, 2012.

[27] Iser, O., Schüren, A. & Schweizer-Ries, P., Bewusstseinsbasierte, transformative Nachhaltigkeitswissenschaft. *Nachhaltigkeit in den Sozialwissenschaften, Theorie und Praxis der Nachhaltigkeit*, ed. W. Leal Filho, Springer Nature, in press.

[28] Schneidewind, U., Singer-Brodowski, M., Augenstein, K. & Stelzer, F., Pledge for a Transformative Science, *Wuppertal Paper*, no. 191, July 2016.

[29] Schneidewind, U., Transformative Wissenschaft – Motor für gute Wissenschaft und lebendige Demokratie. *GAIA*, **24(2)**, pp. 88–91, 2015. https://doi.org/10.14512/gaia.24.2.5

[30] Scharmer, C.O. & Käufer, K., Awareness-based action research: catching social reality creation in flight. *The SAGE Handbook of Action*, ed. H. Bradbury. Sage Publications: London, pp. 199–210, 2015.

[31] Wenger, E., Communities of practice: a brief introduction. *STEP Leadership Workshop*. University of Oregon: Oregon, pp. 1–7, 2011.

[32] Presencing Institute. available at https://www.presencing.org/aboutus/theory-u (accessed 09 August 2021).

[33] Wanner, M., Stelzer Dr., F., Reallabore Perspektiven für ein Forschungsformat im Aufwind. *Wuppertaler Impulse zur Nachhaltigkeit*. Wuppertal Institut: Wuppertal, 2019.

[34] Defila, R. & Di Giulio, A., eds., *Transdisziplinär und transformativ forschen. Eine Methodensammlung*. Springer Nature: Berlin, 2018.

[35] Guided Journaling; Presencing Institute, available at https://www.presencing.org/resource/tools/guided-journaling-desc (accessed 09 August 2021).

[36] Protoyping; Presencing Institute, available at https://www.presencing.org/resource/tools/prototyping-descavailable (accessed 08. August 2021).

[37] Muff, K., *Five Superpowers for Co-Creators. How Change Makers and Business Can Achieve the Sustainable Development Goals*. Routledge: London, 2019.

[38] Fein, E. et al., *LiFT Methodology Book. Designing and Hosting Collaboratories*, 2018, available at http://leadership-for-transition.eu/?page_id=629 (accessed 10 August 2021).

[39] Evangelical Academy Bad Boll, A*rt of Hosting: Die Kunst, Räume für gute Gespräche zu schaffen,* 2016, available at https://christophweinmann.de/aoh-handbuch-bad-boll2016.pdf (accessed 08 August 2021).

[40] Watzlawick, P., et al., *Menschliche Kommunikation*. Hans Huber-Verlag: Bern, 2000.

[41] Schallmo, D.R.A. & Lang, K., *Design Thinking erfolgreich anwenden. So entwickeln Sie in 7 Phasen kundenorientierte Produkte und Dienstleistungen*. Springer Gabler: Wiesbaden, 2020.

[42] Alexandrakis, J., Cycling towards sustainability: the transformative potential of urban design thinking in a sustainable living lab. *Transportation Research Interdisciplinary Perspectives*, **9**, 100269, 2021.

[43] https://www.hochschule-bochum.de/fbe/fachgebiete/energiewendebauen-mondowi/

[44] Schneidewind, U., *Die Große Transformation. Eine Einführung in die Kunst gesell-schaftlichen Wandels*. Fischer: Frankfurt a.M., 2018.

[45] Pardo-del-Val, M. & Martinez-Fuentes, C., Resistance to change: a literature review and empirical study. *Management Decision*, **41(2)**, pp. 148–155, 2003. https://doi.org/10.1108/00251740310457597

PUBLIC VIEWS OF THE VALUE, POTENTIAL, AND SUSTAINABILITY OF ENERGY SOURCES OVER THE LAST 30 YEARS IN THE PACIFIC NORTHWEST, USA

ROBERT L. MAHLER
Department of Soil and Water Systems, University of Idaho, USA.

ABSTRACT

The use of renewable energy has been an important topic in the four Pacific Northwestern states for the last 30 years. Large, statistically designed public surveys were conducted in the region in 1990, 2000, 2010, and 2020 to determine the perceived sustainability, future viability, and acceptance of the following ten energy sources: biomass, coal, geothermal, hydropower, natural gas, nuclear, oil, solar, tidal, and wind power. The survey questions were identical in all 4 years of the survey. These surveys were delivered by the US Postal Service to over 3500 randomly chosen residents in each survey year. The public response rate exceeded 50% in each survey year. Demographic data about age, gender, education level, community size, and state of residence of survey respondents were also collected. The survey data were statistically analyzed. In general, the public was literate identifying the renewable and nonrenewable energy sources as the majority of survey respondents correctly identified biomass, geothermal, hydropower, solar, and wind as renewable energy sources. Based on survey results, over 75% of Pacific Northwest residents considered it important or very important that their energy resources were renewable in 2020. The findings of this study were important because it shows that the public is in line with the scientific community with the goal of greatly reducing energy reliance on C containing nonrenewable energy sources including oil, coal, and natural gas. In summary, (1) the public strongly supports the transformation to a sustainable energy system using primarily renewable energy sources, (2) the use of traditional nonrenewable energy sources like natural gas should not be discouraged in the present; however, they should be phased out over the short and medium terms, (3) solar and wind energy should be significant sources to meet future energy needs in the region, and (4) the renewables including biomass and geothermal have a place in the future energy mix within the Pacific Northwest.
Keywords: hydropower, public opinion, renewable energy, solar energy, sustainable energy, wind energy.

1 BACKGROUND

Climate change has the potential to have a serious negative impact on human civilization. The climate has warmed by over 1.2 °C in the last 220 years. A temperature increase of over 2 °C is predicted to have negative impacts on human health, food production, water supply, and biodiversity. Increasing emissions of CO_2 are largely responsible for this observed temperature change. It is believed that almost 75% of CO_2 emissions are caused by the burning of oil, coal, and natural gas to produce energy for society [1]. Consequently, it is imperative that the energy industry transform itself into a carbon-neutral system in the next 25 years to protect life as we know it on Earth.

2 INTRODUCTION

The threat of catastrophic climate change will require rapid decarbonization of the world's current energy systems making renewable energy sources an important part of the solution to this issue [1]. Compared to coal, oil, and natural gas, nuclear, wind, solar, geothermal, tidal, and biomass power result in low carbon emissions and consequently may be important in the mitigation of the adverse effects of climate change [2]. China and the United States, the two largest sources of global CO_2 emissions, are currently promoting the use of renewables

© 2022 WIT Press, www.witpress.com
DOI: 10.2495/EQ-V7-N1-48-58

including nuclear power as a necessary response to limit global climate change [3]. Many countries have signed on to the Paris Accord that has the primary goal of limiting carbon emissions. Several developed countries including Germany, France, Japan, and the United Kingdom have developed goals to significantly decrease carbon emissions by 2030. Even though the United States has been slow to accept the goals of the Paris Accord, CO_2 emissions have greatly decreased in the last 6 years as energy produced from coal declined because it has become more expensive than other less C emitting energy sources. On the other hand, over 400 additional coal burning power plants have been proposed in just four rapidly growing countries in Asia – China, India, Indonesia, and Vietnam to meet growing energy needs. Many agree that nuclear power is a viable option to control global greenhouse gas emissions; however, future development and utilization of the nuclear option will require both public acceptance and cooperation [4]. In 2019, renewable sources met 22.3% of the world's energy needs. This 22.3% was split between modern renewables (13.3%) and traditional biomass (9%), which include wood, charcoal, straw from fields, and animal dung. Many scientists discount traditional biomass because, although renewable, it may not be sustainable, and it releases CO_2 into the atmosphere. Modern biomass and hydropower production account for upwards of 70% of this renewable, and sustainable energy, while wind, solar, tidal, nuclear, and geothermal energy account for the other 27% of modern renewable energy.

From an energy source standpoint, the four Pacific Northwest states (Alaska, Idaho, Oregon, and Washington) are unique in the United States because a large share of their electricity is furnished by hydropower [3–6]. In fact, Washington, Oregon, and Idaho are responsible for 45% of the United States's hydropower generation capacity. This results in a larger percentage of their energy being renewable than in other regions of the country. Three of these four states (Washington, Oregon, and Idaho) have also invested in other renewables, including wind, solar, biomass, and geothermal. As of 2019, electricity generation in Washington was 60% hydropower, 16% natural gas, 10% nuclear, and 12% other renewables [3]. In Oregon, hydropower, natural gas, and other renewables accounted for 40%, 36%, and 24% of electricity produced, respectively [4]. In Idaho, hydropower, natural gas, and other renewables (wind) accounted for 55%, 16%, and 29% of electricity produced, respectively [5]. Alaska, having the largest oil and natural gas reserves in the United States, was different as these fossil fuels accounted for 67% of the produced electricity while hydropower supplied the remainder [6]. When hydropower and the other renewables (geothermal, biomass, solar, and wind) are combined, the share of electricity provided by renewable sources was 72%, 64%, 84%, and 33% in Washington, Oregon, Idaho, and Alaska, respectively.

The public must be supportive and engaged for the energy infrastructure system across the planet to be successfully converted to a carbonless renewable energy system. The four Pacific Northwestern states of the United States are a region where a significant portion of energy resources are already considered renewable. Consequently, a repeated measure survey instrument of public opinions was developed to determine public views of the value, potential, and sustainability of energy resources in the four Pacific Northwest states in 1988. The survey was designed to do the following: (1) understand the energy sources the public consider the most important for both energy and electricity use, (2) determine the importance the public places in moving toward a carbon-free energy system, (3) learn how the public identifies renewable and nonrenewable energy sources, (4) identify energy sources the public feel are more viable for sustainability, and (5) see if the public energy source conversion time frame is in line with what the scientific community says needs to be done in a timely fashion. The public was asked about the following 10 energy sources: biomass, coal, geothermal, hydroelectric, natural gas, nuclear, oil, tidal, and wind. The mail-based survey was designed to be administered to the public every 10 years from 1990 to 2020. This study summarizes the findings of these survey questions.

3 METHODOLOGY

A survey instrument was developed to determine public views of the value, potential, and sustainability of energy resources in the four Pacific Northwest states in 1988. This survey was sent to 3500 residents in 1990, 2000, 2010, and 2020. The six survey questions in each of the four surveys were as follows:

Q 1. Which of the following energy sources are renewable (sustainable)? Check all that are renewable: *biomass, coal, geothermal, hydroelectricity, natural gas, nuclear, oil, solar, tidal, wind.*

Q 2. What is the most important energy source for all energy uses in the Pacific Northwest? Choose one: *biomass, coal, geothermal, hydroelectricity, natural gas, nuclear, oil (gasoline), solar, tidal, wind.*

Q 3. What is the most important electricity source in the Pacific Northwest? Choose one: *biomass, coal, geothermal, hydroelectricity, natural gas, nuclear, oil, solar, tidal, wind.*

Q 4. How important to you is it that the energy being used in the Pacific Northwest is renewable within 20 years? Choose one of the following: *very important, important, no opinion, not important.*

Q 5. Which of the following energy sources will become more important (viable) in the Pacific Northwest over the next 20 years? Check all that will become more important: *biomass, coal, geothermal, hydroelectricity, natural gas, nuclear, oil, solar, tidal, wind.*

Q 6. Compared to today, which energy sources SHOULD become MUCH more dominant in the Pacific Northwest in the next 20 years? Check all that should become more dominant: *biomass, coal, geothermal, hydroelectricity, natural gas, nuclear, oil, solar, tidal, wind.*

The survey target audience was a representative sample of the 9,500,000 adult residents of Idaho, Oregon, and Washington that live within the four PNW states. In addition, demographic information, including state of residence, community size, gender, age, and educational level, was also collected.

Each survey was developed using the Dillman methodology and was delivered to clientele via the United States Postal Service [9,10]. A sufficient number of completed surveys was the goal to result in a sampling error of 3–5% [11]. The survey process was also designed to receive a completed survey return rate more than 50%. Addresses were obtained from a professional social sciences survey company (SSI, Norwich, CT). Over 3500 surveys were sent out in each mailing event. Four mailings were planned to achieve the 50% return rate. The mailing strategy used was identical to other surveys that had been routinely conducted in the region [12–15]. It only took three mailings to achieve the target return rate of 50% in 1990 and 2000. Conversely, it took four mailings to achieve the 50% return rate in 2010 and 2020.

Survey answers were coded and entered into Microsoft Excel. Missing data were excluded from the analysis. The data were analyzed at two levels using SAS [11]. The first level of analysis generated frequencies, while the second level evaluated the impacts of demographic factors. Significance ($P < 0.05$) to demographic factors was tested using a chi-square distribution [10,11]. Since similar response rates were observed in all survey years, data analysis procedures were identical for each sampling.

4 RESULTS AND DISCUSSION

The survey methodology was designed to be able to compare resident responses over time so that useful information about energy attitudes could be evaluated. Using the mail-based Dillman survey methodology, response rates of 51.3%, 52.5%, 50.9%, and 51.6% were achieved for the surveys conducted in 1990, 2000, 2010, and 2020, respectively. The goal of greater than a 50% response rate was achieved for all surveys, resulting in a sampling error of less than 5%.

When this survey was first initiated in 1990, the population of the four Pacific Northwest states was approximately 9,000,000 [16]. However, by 2020, the region's population had grown to over 14,800,000 [17]. This 64% population increase resulted in the region becoming more urban and more concentrated in communities with more than 100,000 people over the 30-year study period.

There were several instances in this survey study where the demographic factors of gender, age, education level, community size, and state of residence impacted respondent answers. These instances will be discussed in the following sections.

4.1 Energy and electricity sources

The public perceived oil (gasoline), hydropower, and natural gas as the three most important energy sources in the Pacific Northwestern states in 1990, 2000, 2010, and 2020 (Table 1). Compared to the 1990 survey results, the importance of oil and hydropower declined by 2020, while the importance of natural gas as an energy source increased. The public views were very close to the actual energy production data – as oil, natural gas, and hydropower were the most important energy sources. The public view of the decline of the importance of hydropower and consequent increase in the importance of natural gas by 2020 was correct as stagnant hydropower production and rapid population growth resulted in a greater share of energy coming from natural gas. Between 2% and 8% of survey respondents felt that nuclear power was the most important energy source in the region. Conversely, the other six energy sources (biomass, coal, geothermal, solar, tidal, and wind) were never ranked as the most important energy source in the region.

Table 1: The public views of the most important energy sources in the Pacific Northwest based on regional surveys conducted in 1990, 2000, 2010, and 2020.

Energy source	1990	2000	2010	2020	Significance
	--------------------- % -----------------------				
Oil (gasoline)	38	39	37	34	**
Hydroelectricity	31	29	26	24	**
Natural gas	26	29	32	34	**
Nuclear	5	2	5	8	NS
Other	0	0	0	0	
Significance	****	****	****	****	

NS = not significant; **,***, and **** = significant at the 95%, 99%, and 99.9% level of probability, respectively.

The demographic factors of gender, community size, and state of residence impacted survey respondent choices. Males were more likely to view oil as more important than females, while females ranked hydropower and natural gas as more important than males. Residents of communities larger than 100,000 were more likely to consider oil and natural gas more important than respondents in towns with less than 7000 people. Residents of Alaska were more likely to consider oil more important than residents of Idaho, Oregon, and Washington. This was expected since Alaska has the largest oil reserves in the United States, and the state's economy is very oil dependent. The demographic factors of age and education level did not affect respondents' answers to this survey question.

The vast majority of survey respondents identified hydropower as the main source of electricity in 1990, 2000, 2010, and 2020 (Table 2). This observation is correct when compared to the actual generation of electricity data. However, the percentage of respondents that identified hydropower as the major source declined over time. Hydropower was the main electricity source cited by 90%, 89%, 85%, and 77% of the public in 1990, 2000, 2010, and 2020, respectively. The decrease over time is similar to actual generation data and can be attributed to two related factors. The amount of hydropower produced in the region has been relatively stagnant over the last 30 years, while the demand for electricity has increased by over 45% due in large part to an increase in the region's population. The additional electricity has been provided by natural gas. Although not nearly as important as hydropower generation, there has been a significant growth in electrical power generation by both natural gas and wind since 2000. Some consumers noted this as they chose increased natural gas and wind energy production in 2020 and 2010 compared to 1990. The other seven electricity sources (biomass, coal, geothermal, nuclear, oil, solar, and tidal) were never identified by more than 2% of the public as being the major electricity source in the region.

The demographic factors of gender, age, education level, and state of residence impacted how the public chose their major electricity producer in the region. Females were more likely

Table 2: The public views of the most important electricity sources in the Pacific Northwest based on regional surveys conducted in 1990, 2000, 2010, and 2020.

Energy source	1990	2000	2010	2020	Significance
	---------------------- % --------------------				
Hydropower	90	89	85	77	***
Natural gas	6	5	7	10	**
Wind	0	1	3	7	**
Solar	2	2	2	2	NS
Nuclear	1	2	2	3	NS
Geothermal	1	0	0	0	NS
Biomass	1	1	1	1	NS
Tidal	0	0	0	0	NS
Coal	0	0	0	0	NS
Oil	0	0	0	0	NS
Significance	****	****	****	****	

NS = not significant; **,***, and **** = significant at the 95%, 99%, and 99.9% level of probability, respectively.

than males to choose hydropower as the major electricity source than males. Conversely, males were more likely than females to choose natural gas and wind as the major electricity producers. Survey respondents over 50 years old were more likely than younger people to rank hydropower the major electricity source. Residents with a college degree were more likely to choose hydropower as the major electricity source than people with only a high school diploma. Residents of Alaska were less likely to choose hydropower as their major source of electricity than residents of Idaho, Oregon, and Washington. The demographic factor of community size did not affect survey respondent choices.

4.2 What is renewable energy?

Survey year did not affect how residents rated the renewability of 8 of the 10 energy sources evaluated (Table 3). Survey year only impacted the rating of geothermal and wind energy. In general, the public was literate identifying the renewable and nonrenewable energy sources. The majority of survey respondents correctly identified biomass, geothermal, hydropower, solar, and wind as renewable energy sources. Conversely, coal, natural gas, nuclear, and oil were correctly identified as nonrenewable. On the other hand, the public was split about tidal energy – the same controversy is seen in the scientific community where most view tidal as renewable but possibly not sustainable because of adverse impacts on aquatic life. Overall, the public exhibited good literacy in the identification of renewable energy resources.

4.3 Importance of renewable energy

Based on survey results, over 75% of Pacific Northwest residents considered it important or very important that their energy resources were renewable in 2020 (Table 4). The percentage

Table 3: The percentage of surveyed public in 1990, 2000, 2010, and 2020 that considered 10 energy sources as being renewable (sustainable) in the Pacific Northwest. Based on surveys conducted in Alaska, Idaho, Oregon, and Washington.

Energy source	1990	2000	2010	2020	Significance
	------------------renewable, %----------------				
Biomass	50	46	61	55	NS
Coal	6	9	4	4	NS
Geothermal	64	70	79	81	***
Hydropower	86	91	84	80	NS
Natural gas	18	21	16	13	NS
Nuclear	22	26	31	27	NS
Oil	10	4	6	8	NS
Solar	90	93	89	94	NS
Tidal	45	38	43	38	NS
Wind	84	86	90	95	**
Significance	****	****	****	****	

NS = not significant; **, ***, and **** = significant at the 95%, 99%, and 99.9% level of probability, respectively.

Table 4: The importance of energy being renewable (sustainable) in the Pacific Northwest based on Alaska, Idaho, Oregon, and Washington survey data.

Importance	1990	2000	2010	2020	Significance
			%		
Very important	45	53	57	60	***
Important	19	15	13	18	NS
Not important	30	24	20	16	***
No opinion	6	8	10	6	NS
Significance	****	****	****	****	

NS = not significant; *** and **** = significant at the 99% and 99.9% level of probability, respectively.

of survey respondents indicating that it was very important for energy resources to be renewable was 45%, 53%, 57%, and 60% in 1990, 2000, 2010, and 2020, respectively. Conversely, the percentage of the public indicating that it was not important for their energy resources to be renewable declined from 30% in 1990 to only 16% in 2020 (Table 4). These data are important because it shows that the public in this region are willing to remove carbon from the energy generating system and support a renewable energy system. Thus, it is likely that the residents will support both public and private initiatives that will increase the use of both solar and wind renewable energies. This should be relatively straight forward in the case of wind because it is already cost competitive with other widely used energy sources. On the other hand, solar is currently not as cost competitive; however, as this technology improves, it should be embraced in geographic areas in the region with the appropriate amount of sunlight.

All demographic factors impacted how residents answered the renewability question (Table 5). Females were more likely than males to support renewable sources of energy. People younger than 30 years old were most likely to support renewable energy, while residents older than 75 years old were the least likely. People with college education were the most likely to support renewable energy, while residents with 12 or less years of education were least likely to support renewable energy. Support for renewable energy was greatest in communities with more than 100,000 people, while the smallest communities (<3500) had the lowest support level for renewable energy. Oregon and Washington residents were most likely to support renewable energy, while the least amount of support was found with Alaska residents.

4.4 Viability of renewable energy in the region

Based on the data evaluated in the three previous sections, there is good evidence that the public feel the region should convert to primarily renewable energy resources. Currently, nonrenewable energy resources generate over 50% of the region's electricity and approximately 28% of the region's overall energy needs.

The public considers biomass, geothermal, solar, tidal, and wind as energy sources that will become more viable in the region in the next 20 years (Table 6). The public has also

Table 5: The impact of demographic factors on the willingness of people to support the use of renewable energy based on surveys conducted in 1990, 2000, 2010, and 2020.

Demographic factor	Significance	Most pro renewable	Least pro renewable
Gender	***	Female	Male
Age	**	< 30 years old	> 75 years old
Education	**	College	12 years or less
Community size	**	> 100,000 people	< 3500 people
State of residence	****	Washington + Oregon	Alaska

, *, and **** = significant at the 95%, 99%, and 99.9% levels of probability, respectively.

Table 6: The percentage of surveyed public in 1990, 2000, 2010, and 2020 that considered energy sources to become much more viable in the region over the next 20 years in the Pacific Northwest. Based on surveys conducted in Alaska, Idaho, Oregon, and Washington.

Energy source	1990	2000	2010	2020	Significance
	----------------more viable, %----------------				
Biomass	8	15	17	25	**
Coal	2	4	3	1	NS
Geothermal	10	13	13	20	**
Hydropower	65	60	51	38	***
Natural gas	18	16	23	23	NS
Nuclear	16	22	26	21	NS
Oil	12	8	15	17	NS
Solar	35	40	44	46	**
Tidal	2	6	12	15	**
Wind	27	30	37	48	***
Significance	****	****	****	****	

NS = not significant; **, ***, and **** = significant at the 95%, 99%, and 99.9% level of probability, respectively.

viewed these energy sources as more viable in 2020 compared to their feelings about viability in 1990. Based on the data shown in Table 6, the potential viability of solar power has increased from 35% of the public in 1990 to over 46% by the public in 2020. Likewise, the percentage of the public considering wind power as a viable energy source increased from 27% in 1990 to 48% in 2020. Although the public was less confident in the short-term viability of biomass (25%), geothermal (20%), and tidal (15%) energy in 2020, these numbers are substantially higher than they were in 1990, 2000, and 2010 (Table 6).

The demographic factors of gender, age, education level, and state of residence impacted public answers to this question. Females were more likely than males to say that solar and wind energy are viable renewables in the short term. Survey respondents less than 40 years old were more likely than respondents older than 70 to consider solar, wind, and biomass viable energy sources in the short term. Respondents that attended college were more likely to support future viability of renewable energy sources than respondents without exposure to college. Finally, Alaskans were the least likely residents to consider most renewable energy sources viable in the future.

Residents were asked to identify the energy sources that should become much more dominant in the future (Table 7). Note, e.g., hydropower is already a dominant energy source – so relatively few people would say it should become much more dominant. Evaluating the 2020 survey data first, 44%, 35%, 32%, and 29% of respondents said that wind, natural gas, solar, and geothermal energy should become much more dominant over time in the region, respectively. It should be noted that only natural gas is a nonrenewable energy source.

At this point, the public have answered six different questions about energy resources. When the results of these questions are compiled, the following can be said: (1) the public strongly supports the transformation to a sustainable energy system using primarily renewable energy sources, (2) the use of traditional nonrenewable energy sources like natural gas should not be discouraged at the present; however, they should be phased out over the short and medium terms, (3) solar and wind energy should be significant sources to meet future energy needs in the region, and (4) the other renewables including biomass and geothermal have a place in the future energy mix within the Pacific Northwest.

Compared to 1990, residents increasingly said that biomass, geothermal, natural gas, nuclear, solar, tidal, and wind should become more important in the energy mix in 2000, 2010, and 2021. This increasing upward trajectory over time may indicate that people want energy system conversion at a much faster rate than has traditionally occurred over the last 50 years. The demographic factor of gender did have an impact on how people answered this

Table 7: The energy sources that should become much more dominant in the region based on views of the surveyed public in 1990, 2000, 2010, and 2020 in Alaska, Idaho, Oregon, and Washington.

Energy source	1990	2000	2010	2020	Significance
	------------------------ % ------------------------				
Biomass	6	5	11	19	***
Coal	2	6	6	3	NS
Geothermal	6	10	16	29	***
Hydropower	6	10	7	8	NS
Natural gas	10	18	29	35	***
Nuclear	7	12	16	15	**
Oil	4	9	6	10	NS
Solar	18	24	29	32	***
Tidal	1	3	7	5	**
Wind	30	34	38	44	***
Significance	****	****	****	****	

NS = not significant; **, ***, and **** = significant at the 95%, 99%, and 99.9% level of probability, respectively.

question as females were more likely to say that biomass, geothermal, nuclear, solar, tidal, and wind should become more dominant over time than males.

5 CONCLUSIONS AND RECOMMENDATIONS

The major findings of this 30-year survey study were as follows:

- The public perceived oil (gasoline), hydropower, and natural gas as the three most important energy sources in the Pacific Northwestern states in 1990, 2000, 2010, and 2020. These observations are in agreement with actual energy generation data.
- The vast majority of survey respondents identified hydropower as the main source of electricity in 1990, 2000, 2010, and 2020. This observation is in agreement with actual electricity generation data.
- In general, the public was literate in the identification of the renewable and nonrenewable energy sources as the majority of survey respondents correctly identified biomass, geothermal, hydropower, solar, and wind as renewable energy sources.
- Based on survey results, over 75% of Pacific Northwest residents considered it important or very important that their energy resources were renewable in 2020. The percentage of survey respondents indicating that it was very important for energy resources to be renewable was 45%, 53%, 57%, and 60% in 1990, 2000, 2010, and 2020, respectively.
- The public considers biomass, geothermal, solar, tidal, and wind as energy sources that will become more viable in the region in the next 20 years. The public has also viewed these energy sources as more viable in 2020 compared to their feelings about viability in 1990.
- Based on the data evaluated in this paper, there is good evidence that the public feel that the region should convert to primarily renewable energy resources.
- The public are eager to have their energy mix within the region become more renewable over the next 20-year period.

The demographic factors of gender, age, formal education level, state of residence, and community size often affected the response of residents to survey questions. Females were more likely than males to support movement to renewable energy sources. Residents of Idaho, Oregon, and Washington had similar feelings about the need to transition toward a more renewable, sustainable energy environment. Conversely, residents of Alaska, a state rich in natural gas and oil reserves, were less likely to support renewable energy resources. The findings of this study were important because it shows the public is in line with the scientific community goal of greatly reducing energy reliance on C-containing nonrenewable energy sources including oil, coal, and natural gas.

REFERENCES

[1] Dong, K., Dong, X. & Dong, C., Determinants of the global and regional CO_2 emissions: What causes what and where? *Applied Economics*, **51(46)**, pp. 5031–5044, 2019. https://doi.org/10.1080/00036846.2019.1606410

[2] Truelove, H.B. & Greenberg, M., Who has been more open to nuclear power because of climate change? *Climate Change*, **116**, pp. 389–409, 2013. https://doi.org/10.1007/s10584-012-0497-2

[3] REN21, *Renewables 2015: Global status report*. Paris: REN21 Secretariet, 2015. ISBN 978-3-9815934-6-4

[4] Mahler, R.L. & Barber, M.E., University student perceptions of the current and future role of non-carbon emitting energy sources in the world. *International Journal of Energy Production and Management*, **2(3)**, pp. 277–287, 2017. https://doi.org/10.2495/eq-v2-n3-277-287

[5] U.S. Energy Information Administration, *Washington State Energy Profile Overview*, 2021. https://www.eia.gov/washington. Accessed February 18, 2021.

[6] U.S. Energy Information Administration, *Oregon State Energy Profile Overview*, 2021. https://www.eia.gov/oregon. Accessed February 18, 2021.

[7] U.S. Energy Information Administration, *Idaho State Energy Profile Overview*, 2021. https://www.eia.gov/idaho. Accessed February 18, 2021.

[8] U.S. Energy Information Administration, *Alaska State Energy Profile Overview*, 2021. https://www/eia.gov/alaska. Accessed February 18, 2021.

[9] Salent, P. & Dillman, D., *How to Conduct Your own Survey*. John Wiley and Sons, Inc. New York, NY, 1994.

[10] Dillman, D., *Mail and Internet Surveys: The Tailored Design Method*. John Wiley and Sons, Inc. New York, NY, 2000.

[11] SAS Institute Inc., *SAS Online Document 9.1.3*. Cary, North Carolina: SAS Institute Inc., 2004.

[12] Mahler, R.L., Simmons, R., Sorensen, F. & Miner, J.R., Priority water issues in the Pacific Northwest. *Journal of Extension* [On-line], **42(5)**, Article 5RIB3, 2004. Available at: http://www.joe.org/joe/2004october/rb3.php

[13] Mahler, R.L., Simmons, R. & Sorensen, F., Drinking water issues in the Pacific Northwest. *Journal of Extension*, **43(6)**, 6RIB6, 2005. Online at: http://www.joe.org/joe/2005december/rb6.php

[14] Mahler, R.L., Barber, M.E. & Shafii, B., Urban public satisfaction with drinking water since 2002 in the Pacific Northwest, USA. *International Journal of Sustainable Development and Planning*, **10(5)**, pp. 620–634, 2015. https://doi.org/10.2495/sdp-v10-n5-620-634

[15] Mahler, R.L., Barber, M.E. & Simmons, R., Public concerns about water pollution between 2002 and 2017 in the Pacific Northwest, USA. *Int. J. Environ. Impacts*, **2(1)**, pp. 17–26, 2019. https://doi.org/10.2495/ei-v2-n1-17-26

[16] Wikipedia, *List of states of the United States by population*, 2021. Accessed January 2021.

[17] United States Bureau of the Census, *Current population reports, Series P-25, No. 1017. Projection of the population of states by age, sex and race 1988 to 2010*. U. S. Government Printing Office, Washington DC, 1988.

INLAND RAIL FREIGHT SERVICES WITH LESS FUEL AND LOWER EMISSIONS

FRANS BAL[1] & JAAP VLEUGEL[2]
[1] University of Applied Sciences, Utrecht, The Netherlands.
[2] Delft University of Technology, The Netherlands.

ABSTRACT

Many countries have enhanced their air quality agenda (NO_x, PM_x etc.) by a climate change agenda (CO_2 etc.). A direct way to lower these emissions is by using less energy (fuel) per activity. One of these activities is freight transport. Transport from supplier to factory relies on efficient and cost-effective means of transport. Road transport (trucking) is usually preferred. But, trucking is still very dependent on fossil fuels. It is also not suitable for bulk transport over longer distances. In areas without suitable waterways, rail is a logical alternative, but is has its own perils. This paper discusses options to make bulk freight services between Germany and France compliant with emission reduction targets. This leads to the main research question: Is it possible to design rail freight routes that reduce fuel use, emissions of CO_2, NO_x and PM_{10}, while offering competitive transport times? Main rail corridors show signs of congestion and lack of resilience. It is then interesting to research if (dormant) regional/rural, non-electrified, rail tracks could provide capacity and increase resilience of rail services. Such services could also benefit rural economies. A literature study and conversations with a regional expert were used to develop a case study with a rail service using alternative routes. A model was used to estimate the fuel consumption, emissions and trip times of such services. The study indicates that it takes concerted action to achieve the intended goals.
Keywords: cross-border, economics, emissions, Europe, evaluation, Freight transport, logistics, road, rail, policy-making, simulation.

1 INTRODUCTION

Freight transport is a vital industry. It allows producers of goods to connect to suppliers and (final) consumers of their goods. This activity is growing more or less continuously due to changing demands and opportunities [1, 2]:

- A growing world population with an increasing demand for goods;
- A rising average household income, more disposable income and again a rising demand for goods;
- A demand for more diverse and year round availability of goods, which fuels imports from all over the world;
- Producers aiming to optimise cost of production and inventory, which leads to more reliance on just-in-time availability. Stocks are minimized. This means more frequent and smaller shipments. Loading units like containers are less full, while transport vehicles have a lower loading factor and run partially empty, hence more vehicles are needed;
- Cost minimisation also stimulates the creation of supply chains within larger and more complex trade networks. This means more transport over longer distances and resilience risks;
- Both consumers and produces have a preference for short delivery times, which increases demand for rapid, energy-intensive means of transport, in particular air freight and road transport.

1.1 Externalities and policy-making

More freight transport goes along with an increasing fuel consumption and emissions (Fig. 1). Road transport has the largest modal share. As a consequence, it dominates in both areas.

This research involves countries in the European Union (EU) and their options to reduce the CO_2-emissions of freight transport. Its main policy unit, the European Commission, announced an ambitious target of -60% CO_2 for the year 2050 compared to base year 1990, and lower thereafter. Freight transport should achieve these policy targets by a major modal shift from road to rail and inland waterways, and a transition to non-fossil fuels [3].

1.2 Research goal and scope

Rail freight services are well established for medium to longer distances (500 kms and more). They are rarer over shorter distances. Where they exist, they often have to compete with trucking. This paper is about improvements of (existing) bulk train services. It was estimated what the impact of different routes for these services would be in terms of fuel consumption and emissions of CO_2, NO_x and PM_x. This continuation of earlier research [4] integrates the following topics:

- Freight transport and climate change [5] – quantitatively;
- (Re)routing of freight traffic [6] – quantitatively;
- Regional-economic impact of freight transport [7] – qualitatively.

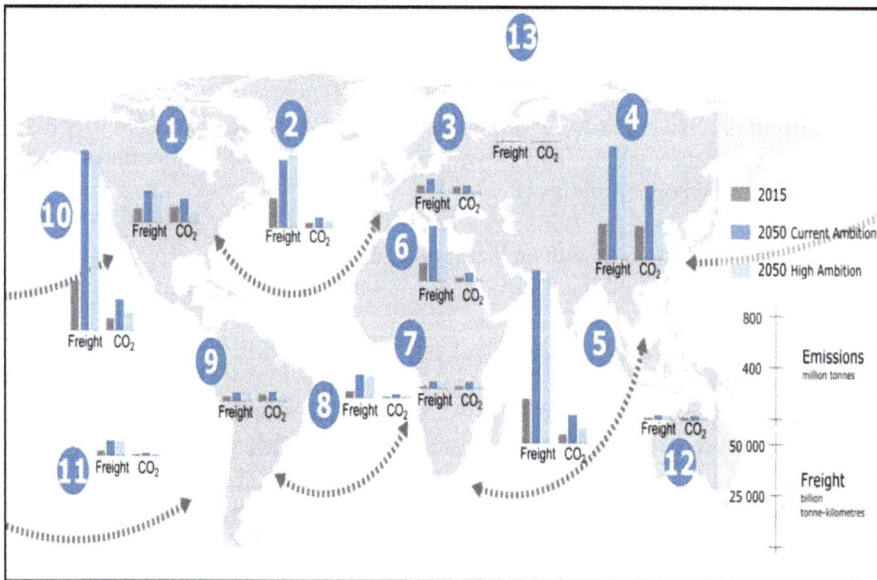

Figure 1: Global freight transport and CO_2-emissions 2015–2050 (Source: [8]).
Legend: 1. North America; 2. North Atlantic; 3. Europe; 4. Asia; 5. Indian Ocean;
6. Mediterranean and Caspian seas; 7. Africa; 8. South Atlantic; 9. Latin America;
10. North Pacific; 11. South Pacific; 12. Oceania; 13. Northern Sea Route

The time horizon is set to the years 2020–2025 to allow minor adaptations of the rail infrastructure and services in the study area.

1.3 Research set-up

The research consisted of fact finding by means of a small literature study and conversations with a regional expert. Relevant data was added to an existing database. A case study was used to design rail service scenarios. These were modeled to estimate the fuel consumption, emissions and trip time for the respective rail. These results were then evaluated and conclusions were drawn.

1.4 Research questions

The main research question is: Is it possible to design rail freight routes that reduce fuel use, emissions of CO_2, NO_x and PM_{10}, while offering competitive transport times?
   The sub-research questions are the following:

- What makes a modal shift a difficult challenge for policy-makers?
- What are interesting freight routes to select for the study?
- What is an interesting example of a product to be transported via this route?
- What are the estimated fuel consumption and CO_2-, NO_x- and PM_{10}-emissions related with such services?
- Is there a time saving if an alternative route would be chosen (comparing main corridor routing with branch line routing)?

2 THE SYSTEM AND THE PROBLEM

2.1 Introduction

Global freight transport is expected to reach triple the 2015 levels by the year 2050 [8]. A doubling of transport is expected for the European Union, a [3]. This will have a sizeable impact on local emissions (NOx and PM_{10}) and greenhouse gas CO_2 (see also Figure 1). Road and (intermodal) rail use and emissions differ substantially (Table 1). The same holds for future projections.
   These figures indicate that the carbon intensity of road and rail differs significantly on a tonkm basis. The difference is smaller in practice due to:

Table 1: Freight transport and CO_2-emissions by road and rail, Europe 2015–2050 (Source: [3].)

	2015	2050	CO_2 business as usual		CO_2 high ambition
			2015	2050	2050
	Tonkms	Tonkms	Mton	Mton	Mton
Road	2819,9	5380,4	161,6	200,7	82,9
Rail	622	1335,4	2,1	1,3	0,2

- Detours. The road network is much denser than the rail network and the number of in-termodal terminals is also limited. This could be solved by infrastructure upgrading and better planning though;
- Use of diesel trucks in intermodal transport. Electric trucks could be used, but their payload and range is still limited, though;
- Emissions are usually stated in terms of a movement with a load factor (fill rate) back and forth. Less payload leads to higher emissions per tonkm and loss of income for the service provider. Cost of the freight service (one way) may also increase. Return payload could be organized, though;
- Freight locomotives are mainly powered by diesel or electricity. This diesel can either be conventional methanol or biomethanol. Electricity can come from mixed sources or green sources. The more green fuel is used, the lower the CO_2-emissions by rail freight services.

The difference in overall emissions by road and rail is then largely explained by the different volumes transported, the fuels used, the total fuel consumption and emission factors of the fuels used to power trucks and freight locomotives.

Barge transport, the second modality in terms of volume in some European countries, is not considered in this paper.

2.2 Policy trade-offs

Freight transport has benefits and costs to society. Benefits include provision of goods, (direct and indirect) employment and tax income. They explain why governments and financial organisations stimulate freight transport with policy instruments like a laws and regulations, investments in infrastructure construction and maintenance [9]. The social costs of freight transport are known as transport externalities. Transport has many of these [10]. The focus will be on two categories:

- Local (NO_x, PM_x, etc.) and global (CO_2) emissions to the air due to combustion of carbon-based fuels in vehicle engines. Such emissions are detrimental for nature, humans and the global climate;
- Consequences of congestion, in itself an internal effect, on main transport corridors. Congestion refers to a (too) small gap between traffic intensity I and available infrastructure capacity C per relevant metric (hour, day). A queing vehicle has an engine that combusts more fuel less efficient and as a consequence produces more emissions. Transport time and cost will also rise in case of congestion.

Governments are in a challenging position when balancing these social benefits and costs. Policy instruments, in particular fuel economy and emission standards, are used to stimulate technical progress. A problem arises when technical progress (a higher fuel efficiency, hence a lower fuel consumption) per (ton)km is partially neutralized by a growing number of vehicles and (ton)kms driven. Then the nett reduction in emissions will be smaller than the technically frontier allows. Additional policy effort is needed to reach the emission target.

If behaviour is the neutralizing factor, then behavioural change should be on the policy-agenda. This can take various forms, including a shift to rail. To make such a shift feasible, several success conditions have to be fulfilled. This starts with clear, transparent and feasible policy targets. For shippers and logistic companies, it should be (made) attractive to use use rail services in the areas where they are represented or do business with.

2.3 Modal split is fairly stable

Today's modal split is the result of developments that took place over many decades. Many regional rail freight services were ended and railway infrastructure was closed (and removed) due to changes in manufacturing and logistics, which reduced the need for bulk transport. This left business with smaller transport volumes (up to a few wagons per time period) a choice between using trucking, leave the area or cease to exist. In most EU countries, freight transport by road has a modal share of at least 75% and it is still growing [11]. If there is potential for a modal shift, then it should be found in individual cases. The idea of a forced general (large) modal shift as promoted by the EU creates a major logistic challenge and may substantially increases transport costs for most shippers and logistic providers.

Rail is mainly focussed on bulk transport (minimum number of wagons; no single wagon load), while trucking is focussed on small volumes per trip (1–2 containers, bins, tanks). Road transport offers fast and frequent delivery at a competitive price. These requirements are hard to meet by rail. Fixed costs of rail are high. There are capacity bottlenecks and gaps in the network, which force detours and longer transport times. The only exception can be found in the main rail corridors, where transport times by road and rail are similar [12], but also there congestion is prevalent. In the European Union (EU), the TEN-T program co-finances studies and projects to "close gaps, remove bottlenecks and technical barriers, as well as to strengthen social, economic and territorial cohesion in the EU". The programme provides small subsidies for studies into and projects to improve infrastructure considered to be of supra-national importance [13].

2.4 (Freight) transport and regional economies

A freight service crossing a region is not necessarily benefical for that region. This holds in particular for international corridors. If the goods transported ar not (off)loaded regionally, then emissions, noise and congestion are not 'off-set' by a regional value-added. This is frequently the reason for public resistance towards (new) infrastructure and a major reason for delay of such projects [14].

Value-added could be generated when a truck or a train departs, stops or ends in the particular region. A region could stimulate this, for instance by opening or enhancing a regional intermodal freight terminal, by adding a missing link or by upgrading a regional branch of a road or rail network. Ideally, this would create a new network allowing transport from and to several destinations. Such a network could have a substantial regional-economic impact. It would help to alleviate the externalities of heavy trucks using regional and local streets, thereby improving liveability. Again, an important condition for feasibility is that shippers and railway operators adapt their logistics.

National and EU policy makers tend to favour the main TEN-T corridors, while the potential of small regional infrastructure receives much less attention. The costs of local upgrades can be limited, but their impact may be important [15]. Improved infrastructure may allow rail operators to develop attractive services. These may then help to preserve or expand regional employment and tax base, allowing pay-back of the investments made.

2.5 Services, capacity and resilience

The rationale for attractive regional rail services not only follows from congestion in main corridors as a consequence of growing traffic, use of terminals in corridors and the (perceived) quality of the main corridors.

The lower density of the rail network compared to the rail network means that disruptions due to technical failures, driver malfunctions, infrastructure maintenance, accidents or natural distasters may have a bigger impact than similar disruptions would have for trucking. There are examples of incidents, which led to many months of traffic rerouting and delays [16]. Shippers are used to just-in-time services. Once these can no longer be guaranteed, second order effects will surface. Apart from damage claims, some shippers may stop using rail. A call for more resilient railways could be answered by developing and maintaining bypasses or shortcuts at the regional level.

Reducing CO_2 by more rail use is challenging. International freight transport has a regional component (services start and end at some point in a region). Currently, trucks dominate regional freight transport. A key success factor for a significant and lasting shift to rail is then how to take care of the logistic needs of regional shippers and receivers of goods.

3 METHODOLOGY

3.1 Assumptions

A micro level analysis was carried out as in [4]. By understanding the micro level it is relatively straightforward and transparent to use a scale factor to simulate macro level results. The opposite approach would carry the risk that the analysis does not arrive at the micro level, hence not provide a meaningful advice to involved stakeholders.

The main actors are regional producers and suppliers of goods, logistic service providers offering road and/or rail services and relevant local and regional governments in the countries involved. They have found common ground due to the growing urgency of climate change mitigation [17]:

- Goods producers (shippers) aim to reduce their CO_2-emissions per tonkm without complicating logistics or increasing operating costs;
- Rail operators optimize fuel consumption and CO_2-emissions per tonkm;
- Governments stimulate rail by alleviating infrastructure bottlenecks and creating the conditions (legal, procedural, financial) that allow running of freight trains through their regions as part of a cross-border rail freight service. This support also helps to establish (not studied) regional and international passenger rail services via the same rail section(s). This may contribute to the bussiness case and lower the resistance towards (renewed) use of such trajectories.

3.2 Scenarios and modeling

In our previous studies simplified networks were used with two nodes – origin - destination pairs. This coarse approach does not allow precise estimations of fuel consumption and emissions of CO_2, NO_x and PM_{10}. Many hours were spent to update an existing decision support model written in MS© Excel© with quite detailed routes. It allows to simulate various train parameters (freight volume, load factor, locomotive types, train length, train weight), stops and routes.

Technological parameters can be adapted if necessary to simulate technical progress. This includes engine technology, energy category (diesel, alternative), fuel consumption per tonkm, fuel specification (fuel-blend), emission factors and electricity mix (% green) etc.

Only tank-to-wheel emissions were considered. Fuel consumption was taken from the literature and validated by an expert. The impact of a different load factor on fuel consumption and emissions was also added to the model.

4 CASE STUDY

4.1 Introduction

It is interesting to explore the feasibility and benefits of embedding non-main corridor rail sections into a cross-border service as a means to offer attractive, fuel efficient and resilient rail freight services.

The background study included the German Saarland and Rheinland-Pfalz, the northern part of France, Luxemburg, Belgium and the Netherlands. The paper focusses on a small section of this search area.

4.2 Node Dillingen

Dillingen (Germany, Fig. 2) is in the heart of several European road and rail corridors, in particular Atlantic Rail Freight Corridors (RFC) 2 and 4 (Saar Railway Saarbrucken-Trier) (Fig. 3).

These corridors give access to intermodal terminals like those in Bettembourg (Luxembourg) and Dijon-Gevrey (France). An important logistic hub at Ludwigshaven (Germany) is at reasonable distance north-east of Dillingen. The area Saarbrucken-Trier-Metz-Luxembourg has a significant industrial importance. In Germany there are plants like NEMAK

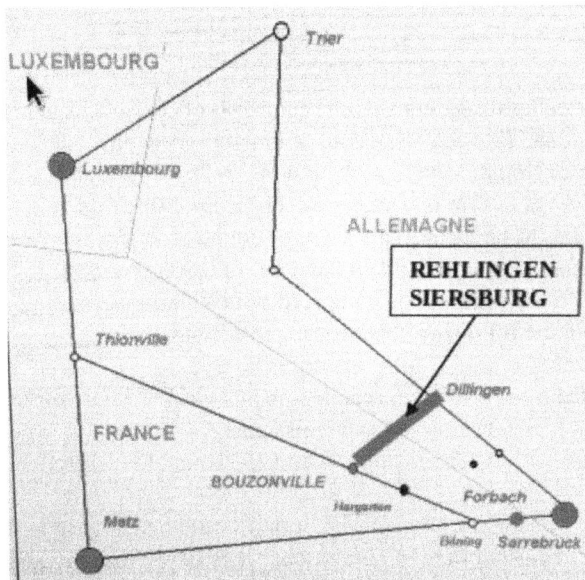

Figure 2: Node Dillingen with missing link (Source: [18]).

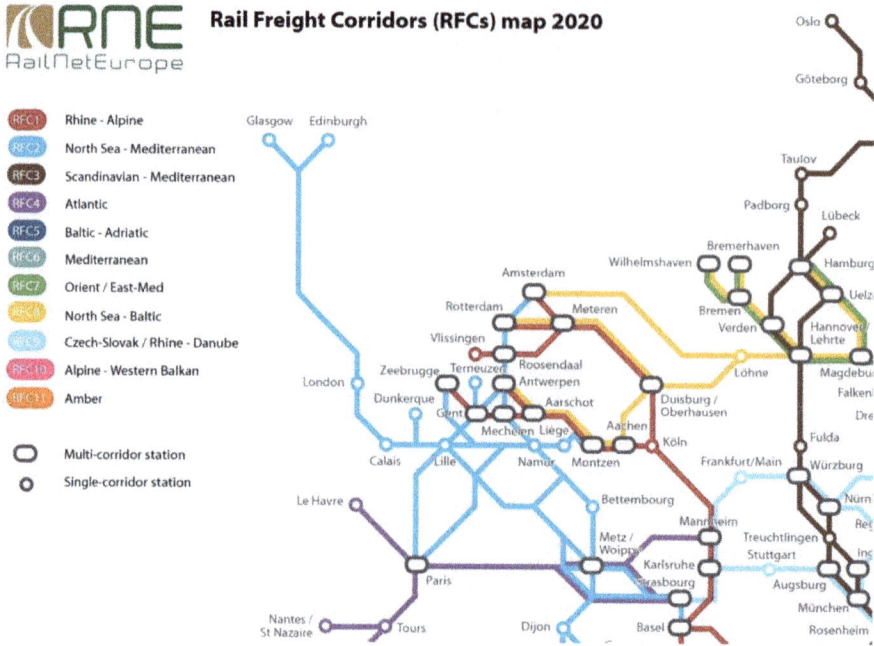

Figure 3: North-western RNE rail corridors 2020 (Source: [19]).

(aluminium), Röderberg Ford Werke (car assembly) and Dillinger Hütte (steel mill). In France there is Total Petrochemicals near Carling / Saint-Avold.

4.3 Chalk transport

The Dillingerhütte steel mill uses considerable amounts of chalk. This is excavated in Dugny-sur-Meuse near Verdun, France. The about 7 km line from the quorrie to Verdun is not electrified. An average freight train has a length of 17–20 wagons and weighs between 1200 and 1500 tonnes. On flat terrain one single diesel locomotive could suffice. However, some significant slopes have to be taken. This asks for double traction; two Vossloh DE18 diesel locomotives. Their fuel consumption is in the order of 0,035 liter/tkm.

The conventional route (Table 2) is compared with a shortcut via the Niedtalbahn (Table 3). This should fulfill the following conditions:

- Allow the intended freight volume, train length and weight;
- Preferable reduce fuel consumption and emissions;
- Allow connections with potential shippers and receivers of bulk products.

Comparing Table 3 and 4, the Niedtalbahn option would result in a 52 km shorter train trip and, hence less fuel consumption and emissions. It is likely that the shorter distance also reduces trip time, but this can only be verified when the real service times (including delays and waiting at terminals) are known.

Table 2: Route 1 – without Niedtalbahn.

Route: Dugnu-sur-Meuse (F) – Verdun (F) – Conflans en Jarnisy (F) – Hagondange (F) – Metz (F) – Forbach (F) – Saarbrücken (G) - Dillingen (G)		Full	Empty
Distance in km	199		
Diesel fuel in ltr		10503	3490
CO_2 in ton		27.83	9.2
NO_x in kg		116	38
PM_{10} in kg		27	0.9

Source: Own estimations

Table 3: Route 2 – with Niedtalbahn.

Route: Dugny-sur-Meuse (F) – Verdun (F) – Conflans en Jarnisy (F) – Hagondange (F) – Thionville (F) – Bouzonville (F) – Dillingen (G)		Full	Empty
Distance in km	147		
Diesel fuel in ltr		7722	2565
CO_2 in ton		20.5	6.8
NO_x in kg		84.9	28.2
PM_{10} in kg		19.9	0.6

Source: Own estimations

4.4 Renovation of the Niedtalbahn

Dillingen-Bouzonville was part of a strategic railway line in the German-French border area. It was inaugurated in 1901. Passenger services stopped in 1945. A service to the yearly Good Friday market in Bouzonville remains. Dillingerhüte receives trains carrying chalk and used to have outbound trains with liquid iron to France. Cross-border services stopped due to German and French decisions. Deutsche Bahn even wanted to remove a bridge, which was not granted. France does not want to pay for a local dispatcher. Rail infrastructure in France needs replacement. Deutsche Bahn renovated the German section. It is still running a passenger service until Niedaltdorf [20]. Politicians in Saarland and Bouzonville favour a reopening, but is not local politicians who decide on this, but higher level authorities, including rail network managers in both countries. Cost-benefit analyses have been carried out, but their outcome is very much dependent on the assumptions made, including dual use by passenger and freight [21].

5 CONCLUSIONS

The case study shows that it is possible to design (alternative) cross-border rail freight routes that significantly reduce fuel use, emissions of CO_2, NO_x and PM_{10}. The proposed bypass may offer a shorter transport time and add to network resilience. A shift to rail is feasible in individual cases, but this means that the right conditions have to be created. This is particularly challenging in a case like this, because of the many stakeholders with diverging interests, the (initially) moderate to low traffic volumes and necessary investments in renewal of railway infrastructure like tracks, tunnels and signalling.

ACKNOWLEDGEMENT

We are grateful for the generous help provided by Herr Erhard Pitzius of the Plattform Mobilität SaarLorLux e.V., Überherrn, Germany. He provided us with many reports and regional information relevant for our study.

REFERENCES

[1] International Transport Forum/OECD, The Carbon Footprint of Global Trade, Tackling emissions form International Freight Transport, http://www.itf-oecd.org/sites/default/files/docs/cop-pdf-06.pdf, Paris, 2015.

[2] Hesse, M. & Rodrigue, J.-P., The transport geography of logistics and freight distribution, *Journal of Transport Geography*, **12(3)**, pp. 171–184, 2004. https://doi.org/10.1016/j.jtrangeo.2003.12.004

[3] European Commission, A European Strategy for Low-Emission Mobility, COM(2016) 501 final, Brussels, 20.7.2016.

[4] Bal., F. & Vleugel, J.M., Towards more environmentally sustainable intercontinental freight transport. *International Journal of Transport, Development and Integration*, **4(2)**, pp. 129–141, 2020. Vleugel, J.M., Bal, F., Regional goods delivery: How to reduce its CO_2-, NO_x- and PM_{10}-emissions? *International Journal of Energy Production and Management,* **3(4)**, pp. 338–347, *2018.* Janic, M., Vleugel, J.M., 2012. Estimating potential reductions in externalities from rail–road substitution in Trans-European freight transport corridors, *Transportation Research. Part D: Transport & Environment,* **17(2)**, pp. 154–160, 2012.

[5] Chapman, L., Transport and climate change: A review. *Journal of Transport Geography*, **15**, pp. 354–367, 2007. https://doi.org/10.1016/j.jtrangeo.2006.11.008

[6] Uddin, M. & Huynh, N., Reliable Routing of Road-Rail Intermodal Freight under Uncertainty. *Netw Spat Econ,* **19**, pp. 929–952, 2019. https://doi.org/10.1007/s11067-018-9438-6

[7] Rodrigue, J.-P., *The geography of transport systems*, 5th ed., Routledge, New York, 2020.

[8] International Transport Forum/OECD, Transport Outlook 2019, http://doi.org/10.1787/transp_outlook-en-2019-en, Paris. Accessed on: 11 April 2021.

[9] See reference 7.

[10] Demir, E., Huang, Y., Scholts, S. & Woensel, T. van, A selected review on the negative externalities of the freight transportation: Modeling and pricing. *Transportation Research Part E: Logistics and Transportation Review*, **77**, pp. 95–114, 2015. https://doi.org/10.1016/j.tre.2015.02.020

[11] Pastori, E., Brambilla, M., Maffii, S., Vergnani, R., Gualandi, E. & Dani, E., Research fro TRAN Committee – Modal shift in European transport: A way forward, research for the European Parliament, http://www.europarl.Europa.eu/RegData/etudes/STUD/2018/629182/IPOL_STU(2018)629182_EN.pdf, retrieved May 2, 2021.

[12] European Court of Auditors, Rail freight transport in the EU: still nog on the right track http://www.eca.europa.eu/Lists/ ECADocuments/SR16_08/ SR_RAIL FREIGHT EN.pdf, retrieved 18 April, 2021.

[13] European Parliament and Council, Regulation (EU) No. 1315/2013 on Union guidelines for the development of the trans-European transport network and repealing Decision No. 661/2010/EU, 2010, retrieved 5 May, 2021.

[14] NIMBY, http://en.wikipedia.org/wiki/NIMBY, retrieved 6 May, 2021.

[15] Vleugel, J. & Bal, F., Some approaches to reduce transport time of intermodal services: Smart rail investments, *European Transport \ Trasporti Europei*, **52(3)**, pp. 1–15, 2012.

[16] An avalanche split a hill with tracks near Kestert, Germany, along the Rotterdam-Genua corridor; the busiest in Europe. It took 1,5 months to restore the tracks. Freight and passenger trains were rerouted, delayed or cancelled. http://www.swr.de/swraktuell/rheinland-pfalz/koblenz/bahnl aerm-mittelrheintal-kestert-felsrutsch-100. html, retrieved 7 May, 2021.

[17] Rice, D., Global temperatures could pass limit set by Paris Climate deal within 5 years. www.usatoday.com/story/weather/2018/02/01/global-temperatures-could-pass-limit-set-paris-climate-deal-within-5-years/1087326001, 1 February, 2018, retrieved 9 May 2021.

[18] Aubin, B., Étude « Bouzonville en train » Proposée par la Ville de Bouzonville et Bernard AUBIN, Fédération Indépendante du Rail et des Syndicats des Transports, 1998.

[19] Rail Net Europe, http://rne.eu/rail-freight-corridors, retrieved 22 May 2021.

[20] Warscheid, L., Neue Chance für Erhalt der Niedtalbahn, Saarbrücker Zeitung, 17 April 2018.

[21] Leyes, J., Wiederaufnahme eines grenzüberschreitenden Schienenverkehres über die Niedtalbahn -Analyse der Chancen und Möglichkeiten, sowie der Kosten und Nutzen, BSc thesis Hochschule Kaiserslautern, 01-02-2021.

ENERGY EFFICIENCY IN INDUSTRY 4.0: ASSESSING THE POTENTIAL OF INDUSTRY 4.0 TO ACHIEVE 2030 DECARBONISATION TARGETS

SIMONE MAGGIORE, ANNA REALINI, CLAUDIO ZAGANO, FRANCESCA BAZZOCCHI, ELENA GOBBI & MARCO BORGARELLO
Ricerca sul Sistema Energetico – RSE S.p.A., Italy.

ABSTRACT
The energy transition for the industrial sector is not limited to a reduction in energy consumption: the real issue is to combine sustainability with growth, by mixing the two ingredients (the rational energy use and the industrial growth) which are not always compatible. The National Energy and Climate Plan (NECP) and the New Green Deal policies in Italy have the goal to promote an economic development as well as the environment sustainability and social inclusion. RSE[1] has investigated the role of the national incentive plan 'Impresa 4.0' in Italy (currently 'Transizione 4.0', equivalent to 'Industry 4.0') as a measure to promote the energy transition, analysing whether and how is it possible to combine economic development with energy efficiency. Originally, it was developed to increase the competitiveness of industrial sector, but, progressively, it was also used to promote energy efficiency and sustainability. A survey was carried out by RSE on about 300 companies that implemented innovation and digitalisation interventions, monitoring the effects and impacts that the '4.0 choice' has determined on energy consumption, on their environmental externalities and, in general, on other costs. Moreover, some case studies were collected, together with a database of 'Impresa 4.0' application, which supported technical and economic evaluations. The impact of these measures on energy performance of the companies was estimated from the analysis of actual projects and from interviews and discussions with the operators. In this paper, the results of the survey are presented and the outcomes are analysed in comparison with the Italian manufacturing sector performance, in order to establish the potential of 'Impresa 4.0' policies in supporting the decarbonisation process and reaching 2030 environmental targets.
Keywords: decarbonisation, energy efficiency, environmental externalities, Industry 4.0.

1 INTRODUCTION
The technological transformation, which is identified with the term 'Industry 4.0', is substantially guided by the intertwining of the processes of technological innovation and digitalisation, to optimise and make production processes more flexible and, consequently, obtain advantages in terms of competitiveness on the market.

In this context, one of the key factors is the energy consumption which must be optimised to achieve production objectives in the least expensive and most profitable way, balancing the several variables involved in production processes. This approach, supported by digital opportunities, allows, in many cases, to reduce the energy intensity of production processes, in the logic and spirit at the basis of the energy transition. To have a deeper perception of the interweaving of innovation/digitalisation and energy efficiency, RSE has planned an in-depth study, making use of the experience and the suggestions of various stakeholders in the sector [1].

The Industry 4.0 incentive plan in Italy was initially developed to increase the competitiveness of the industrial sector; but, progressively, it was reshaped to promote and direct this progress at energy efficiency and sustainability, in order to become a measure to promote the

---

[1] RSE stands for 'Ricerca sul Sistema Energetico' (http://www.rse-web.it/home.page), a publicly owned company whose mission is to carry out publicly funded national and international programmes in the fields of electrical power, energy and the environment.

© 2022 WIT Press, www.witpress.com
DOI: 10.2495/EQ-V6-N4-371-381

energy transition. A new approach to 'efficiency' was created: not only 'energy efficiency' and 'energy saving' but also 'overall system efficiency'. Which means further increase in energy savings and an 'optimisation' of production, environmental constraints, water and soil consumption, and finally also personnel safety.

2 THE SURVEY

2.1 Literature review and previous works

Industry 4.0 represents a great opportunity for companies but, at the same time, it is big challenge which requires new skills and visions to fully obtain its benefits.

It is not surprising, therefore, that Industry 4.0 and the related innovations and potential competitive advantages for companies have undergone and are currently undergoing a profound investigation by many actors, both within the companies and external to them, under the pressure of national government, which are eager to know the results of their policies and put in place the requires corrections in order to strengthen the support mechanisms.

Many of the existing surveys, however, fail to provide quantitative data, in particular on the relation between energy efficiency and Industry 4.0-related technologies and investments. For example, in [2], the authors have investigated how much the Italian manufacturing companies are ready to be concretely involved in the Industry 4.0 journey, focusing on analysing the knowledge and adoption levels of specific technologies, and pointing out the main benefits and obstacles, but no quantitative data are provided. In [3], the authors provide a comprehensive review of Industry 4.0-associated topics such as intelligent manufacturing, Internet of Things (IoT)-enabled manufacturing and cloud manufacturing, but a quantification on the benefits of these items on the companies' business and energy performances cannot be found. [4] shows how Industry 4.0 can be a game changer, helping companies to control their energy spending, allow them to use equipment more efficiently and provide greater flexibility in manufacturing, but, once again, a quantification of such benefits is lacking. [5] states that Industry 4.0 will enable companies not only to organise their production process more efficiently but also to manufacture customised products within the framework of and at the same cost as automated manufacturing; however, at least an estimation of the amount of energy efficiency improvement, and consequent energy and costs savings, is missing. Finally, in the 'Digital Energy Efficiency Report 2020' [6] made by the Energy Strategy Group from Politecnico di Milano, one of the most influential institutions in Italy, quantitative data about the impact of Industry 4.0 technologies in terms of improved energy efficiency in companies, is not available.

Quantitative data would be very helpful to concretely grasp the effect and consequences of Industry 4.0 interventions of the companies' daily activities and business.

2.2 Survey design

As already said before, most of the existing surveys about Industry 4.0 are qualitative; to fill this gap, RSE has planned a survey with the goal to provide quantitative results, using a bottom-up approach starting with companies' experience in the field, thus allowing to draw conclusions on the effectiveness of the national incentive plan 'Impresa 4.0' in Italy using quantitative data provided by companies which actually implemented real intervention in their daily activities.

Thanks to the collaboration with large- and medium-sized companies, with trade associations such as Confindustria, with research institutions such as Cesisp, which have strong links with the University of Milano-Bicocca in Milan, and with service providers such as FIRE, which also manages the register of energy managers in Italy, RSE was granted access to a network of companies which are usually not accessible without having the previously mentioned 'ambassadors' to gain companies' managers' trust and consequent willingness to share sensitive data about investment within their companies.

This monitoring activity has involved the companies that have carried out energy efficiency interventions and/or have made use of the incentives provided for by the Industry 4.0 incentive plan (and subsequent ones) established by the Italian government.

Through the implemented survey, RSE aimed to analyse whether and how it is possible to combine digitalisation and economic development with energy efficiency. Therefore, the respondents were asked to provide information and data related to the impact of the adoption of technologies supported by the Industry 4.0 incentive plan. The collected information included the types of adopted technologies, electricity, heat and water consumption data before and after interventions and related savings. The reduction/increase in waste production and labour costs was also investigated. The impacts were calculated in terms of energy costs and overall production efficiency of companies. It important to point out that, in order to distil only the benefits coming from Industry 4.0 investments, all other factors (such as production increase/decrease, …) have been taken into account.

The reasons and barriers underlying the choice to use the Industry 4.0 incentive plan have been investigated too. The answers to the survey were collected through a Computer-Assisted Web Interview (CAWI) and processed in an anonymous way, as most of them were sensitive data; this has allowed RSE to collect an important series of in-field experiences by those stakeholders operating the industrial sector.

2.3 Characterisation of the sample of companies

The survey was carried out on a sample of companies belonging to the manufacturing sector in Italy, chosen to be as heterogeneous as possible, in order to obtain representative data of industries of different sectors and size. The analysis focused mainly on manufacturing companies as such sector made more use of the Industry 4.0 incentive plan, as also emerges from the report of the Centro Studi Confindustria 'Un cambio di paradigma per l'industria italiana: gli scenari di politica economica' [7] which reports the data, referring to 2017, of the policies for economic growth, including the Industry 4.0 incentive plan. The companies involved in the survey are in the North (65%), in the Centre (24%) and in the South (11%), as shown in Fig. 1.

As shown in Fig. 2, the composition of the sample is characterised for 40% by medium/large companies, with a total revenue larger than 100 million euros and more than 250 employees: therefore, these companies are structurally prepared, for budget availability, size of plants and size of consumption, to perform interventions affecting energy saving and the production process. The remaining part is composed of companies with more than 50 employees (31%) and with less than 50 employees and a turnover of less than 10 million euros (29%).

The companies involved in the survey are, on average, more energy intensive than the national average: 13% of them, in fact, declare a ratio between energy costs and turnover comprised in the 10–15% range, while, the entire national manufacturing sector is

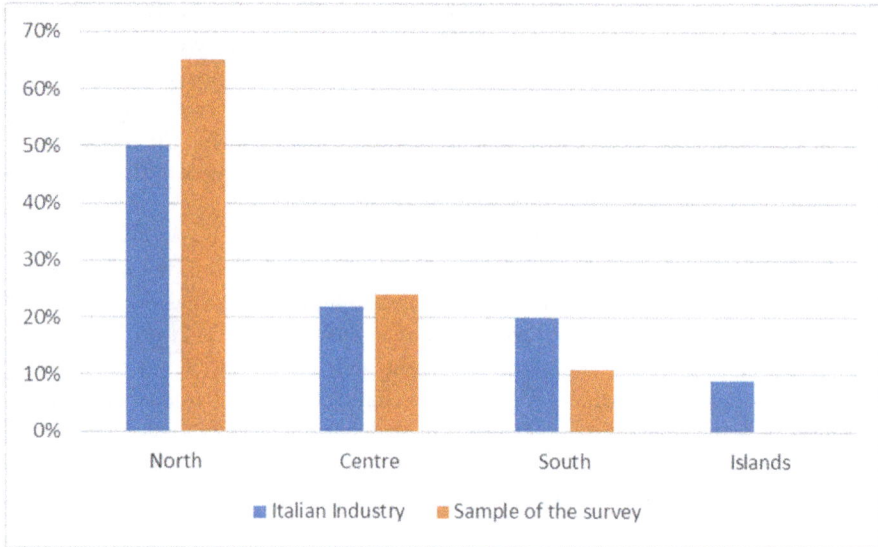

Figure 1: Territorial distribution of companies involved in the survey in comparison to the Italian companies' distribution [1].

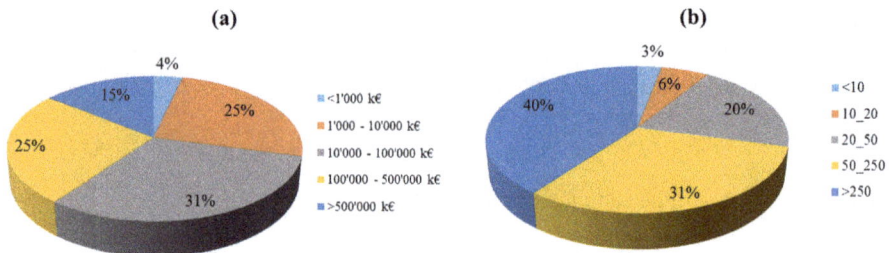

Figure 2: Distribution of companies involved in the survey based on the value of total annual revenues (a) and on the number of employees (b).

considered, this figure goes down to 6% [8]; on the other hand, 40% of the companies involved in the survey can be found in the 0–5% range, while the same figure for the entire national manufacturing sector is 58% [8].

In terms of product coverage, as shown in Table 1 (breakdown by NACE sector classification), 23% of companies belong to the metallurgy sector, that, also according to the Confindustria report [7], is the most active in investing in digital technologies. The following sectors are the processing of non-metallic minerals, in particular glass, ceramic and cement (15%), and the chemical sector (11%).

3 RESULTS OF THE SURVEY

3.1 Energy efficiency interventions

The results of the survey show that 96% of the companies carried out at least one energy efficiency intervention in the last 10 years. Among these, the most widespread intervention

Table 1: Distribution of companies involved in the survey based on NACE sector classification.

NACE	Sector	Percentage
24	Manufacture of basic metals	23.1%
23	Manufacture of other non-metallic mineral products	15.4%
20	Manufacture of chemicals and chemical products	11.3%
	Non-manufacture	10.5%
22	Manufacture of rubber and plastic products	6.9%
25	Manufacture of fabricated metal products, except machinery and equipment	6.5%
17	Manufacture of paper and paper products	5.3%
26	Manufacture of computer, electronic and optical products	4.0%
28	Manufacture of machinery and equipment n.e.c.	4.0%
27	Manufacture of electrical equipment	3.6%
18	Printing and reproduction of recorded media	2.4%
29	Manufacture of motor vehicles, trailers and semi-trailers	1.6%
10	Manufacture of food products	1.2%
19	Manufacture of coke and refined petroleum products	1.2%
31	Manufacture of furniture	1.2%
11	Manufacture of beverages	0.4%
13	Manufacture of textiles	0.4%
21	Manufacture of basic pharmaceutical products and pharmaceutical preparations	0.4%
30	Manufacture of other transport equipment	0.4%

concerns lighting, followed by the installation of inverters or high-efficiency electric motors and from interventions on the compressors, as shown in Fig. 3.

Overall, the interventions resulted in different levels of energy savings depending on the adopted technology. About 60% of the companies involved in the survey declares electrical and/or thermal savings in the 1–5% range, while for the remaining companies, the savings are more significant and can even exceed 10% (Fig. 4).

In 3% of the companies, a 'cross effect' occurred for certain types of interventions: such intervention aimed at reducing only electrical or thermal consumption, but it had the effect of increase the consumption of the other source, although with an overall positive energy balance. This effect often depends on the type of intervention carried out, such as in cases of installation of a system of cogeneration, which can result in an increase in the overall efficiency of the system in the face of a local increase in gas consumption.

3.2 Industry 4.0 interventions

The results of the survey show that 57% of the companies involved in the survey implemented interventions which are eligible for incentives under the Industry 4.0 incentive plan. This shows the interest from companies towards this type of technology and related mechanisms, considering that, with respect to the incentives for energy efficiency (active since

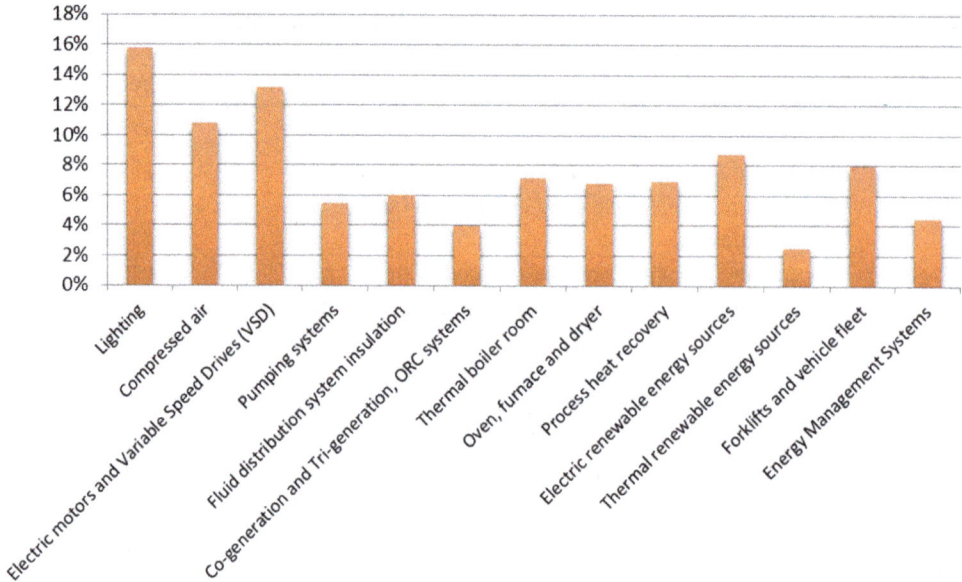

Figure 3: Distribution of energy efficiency interventions among the companies involved in the survey.

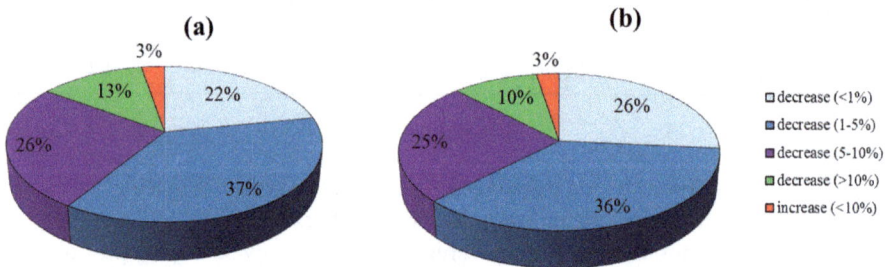

Figure4: Impact of the energy efficiency interventions on the electric (a) and thermal (b) consumption for the companies involved in the survey.

2005), actions to promote Industry 4.0 measures have been introduced only 2017. The interventions were grouped into the following categories [9]:

1. Industrial Internet of Things (IIoT)
2. Cloud Manufacturing
3. Industrial Analytics
4. Advanced HMI
5. Advanced Automation
6. Additive Manufacturing
7. Augmented Reality

8. Advanced Manufacturing Solutions
9. Cybersecurity
10. Horizontal/Vertical Integration
11. Simulation

As shown in Fig. 5, the most applied type of intervention is 'Advanced Automation' (18%), followed by 'Cybersecurity' (13%) and 'Industrial IoT' (11%) interventions.

In order to analyse the impact of Industry 4.0 interventions, RSE has analysed the energy consumption before and after the interventions, thus setting a baseline consumption upon which a reduction/increase could be asserted. Moreover, as already said before, the energy savings were cleaned from other factors affecting the processes independently of Industry 4.0 interventions, such as production increase/decrease or increase in energy costs.

As regards the impact of Industry 4.0 interventions electricity consumption, it results that in more than half of the involved companies, energy savings were measured and such savings are in the range 1–20% for about 50% of the companies. On the contrary, electricity consumption increased as a result of the implementation of Industry 4.0 technologies for 11% of the companies, even if such an increase is less than 5% (Fig. 6).

The reason lies in the larger demand for electricity of some control or automation systems, but also from the different management of the system, that can lead to maximise production at the expenses of energy efficiency.

The impact of Industry 4.0 interventions on thermal consumption is less than the one on electricity consumption: in fact, only 38% of the companies experienced a decrease in thermal consumption; on the other hand, 2% of the companies reported an increase in thermal consumption, attributable, however, to particular cases, relating to specific types of production (Fig. 7).

It is important to highlight that, in this case, energy consumption decrease is a secondary effect and not the reason which led companies to implement a particular Industry 4.0 technology.

In terms of reduction of water and waste consumption, the impact of Industry 4.0 interventions is lower, compared to that of electrical and thermal consumption: in fact, it occurs only in 17% and 21% of companies as regards, respectively, water and waste consumption.

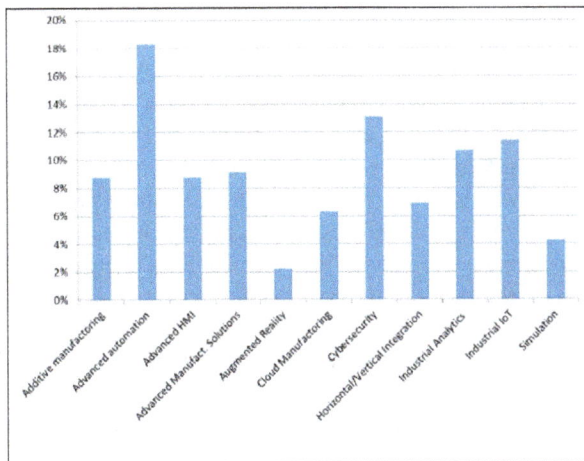

Figure 5: Share of implementation of interventions which are eligible for incentives under the Industry 4.0 incentive plan for the companies involved in the survey.

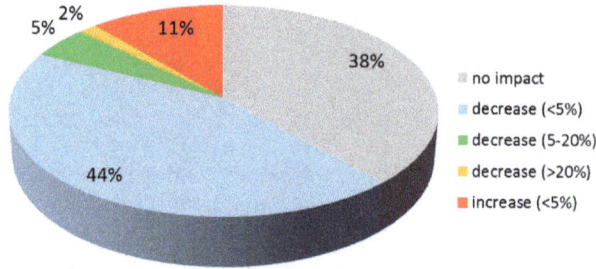

Figure 6: Impact of the Industry 4.0 interventions on the electric consumption for the companies involved in the survey.

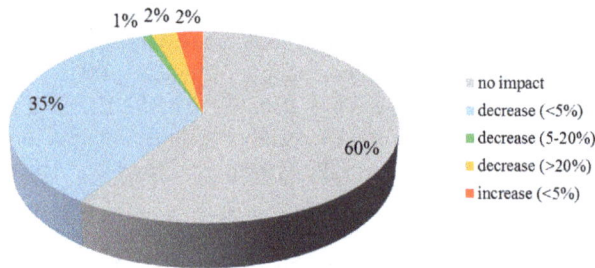

Figure 7: Impact of the Industry 4.0 interventions on the thermal consumption for the companies involved in the survey.

The effect on labour costs, on the other hand, appears to be more significant. In 50% of cases, in fact, digitalisation and automation technologies also led to a reduction in costs of the personnel involved. Eight percent of the companies have, instead, experienced an increase in labour costs, probably attributable to the cost of specific training for the use or the maintenance of the installed technologies (Fig. 8).

Figure 9 shows the average electrical and thermal energy savings for the different Industry 4.0 interventions. The average energy savings can be found in the 0–3% range and are larger in those interventions which provide an organic and integrated vision of the whole production process, such as the 'Cloud Manufacturing' and the exchange of integrated information. Considerable savings can also be generated by those interventions involving a human–machine interface, to convey and organise the information regarding components and process data. It was also noted that, for some interventions, the percentages of energy savings are unbalanced towards thermal ones, i.e. the use of heat for production processes. The capability to interconnect the different phases of the process and the dynamic control of the system components can, in fact, facilitate a more optimised use of thermal vectors, both cold and heat. One of the main reasons for the energy savings achievable by digitisation is the reduction of manufacturing defects. In fact, digitisation allows for more efficient control of processes, resulting in the reduction of defective products. Each avoided defective product translates, in turn, into lower energy consumption and lower consumption of raw materials with consequent economic advantage.

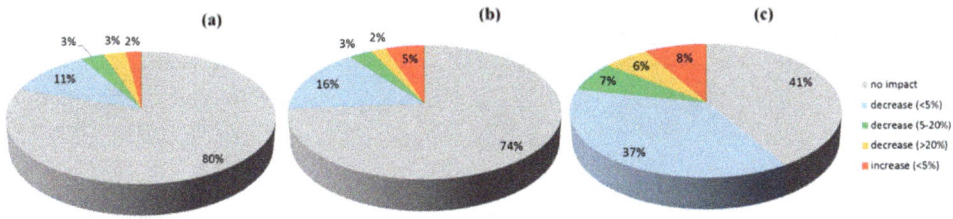

Figure 8: Impact of the Industry 4.0 interventions on the water consumption (a), waste consumption (b) and labour cost (c) for the companies involved in the survey.

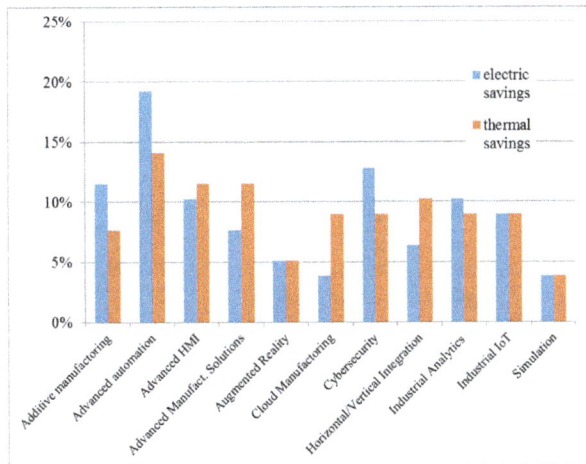

Figure 9: Average electrical and thermal energy savings for the different Industry 4.0 interventions implemented by the companies involved in the survey.

4 ESTIMATION OF POTENTIAL ENERGY SAVINGS FOR THE DIFFERENT SECTORS

Starting from the cross among the data on the types of companies, the implemented Industry 4.0 interventions and the related electricity and thermal savings, electrical and thermal energy savings, with respect to baseline consumption, were estimated, for those NACE sectors with a statistical significance, in order to be able to generalise the answers (Table 2).

Energy savings are in the 0–3% range: this is an estimate based on data from a sample of companies, but, based on RSE experience and knowledge of the industrial sector in Italy, it appears that these values represent a semi-quantitative indication of the savings achievable by Industry 4.0 measures.

5 CONCLUSIONS

The main conclusion of this analysis is that Industry 4.0 incentive plan, even if it does not have the achievement of energy efficiency in companies among its objectives (in fact, many of the companies did not plan Industry 4.0 intervention to increase energy efficiency in their

Table 2: Estimation of potential electrical and thermal energy savings with respect to baseline consumption for some NACE sectors.

NACE	Sector	Potential electric energy savings	Potential thermal energy savings
17	Manufacture of paper and paper products	0.2%	0.2%
18	Printing and reproduction of recorded media		
20	Manufacture of chemicals and chemical products	0.5%	0.3%
21	Manufacture of basic pharmaceutical products and pharmaceutical preparations		
22	Manufacture of rubber and plastic products	2.2%	1.3%
23	Manufacture of other non-metallic mineral products	2.6%	2.2%
24	Manufacture of basic metals	2.9%	2.2%
25	Manufacture of fabricated metal products, except machinery and equipment	3.1%	0.3%
26	Manufacture of computer, electronic and optical products	0	0.3%
27	Manufacture of electrical equipment		
28	Manufacture of machinery and equipment n.e.c.	1.5%	0

companies but to improve their digitalisation), was able to improve it in many of the companies who benefitted from such a support scheme.

In fact, the constant monitoring and control of machinery equipped with technologies allowing process optimisations were able to improve energy efficiency in companies, leading to a reduction not only in electricity and thermal consumptions but also, in some cases, in water and waste consumptions and in labour costs.

This digitalisation and related benefits stemming from Industry 4.0 intervention will certainly make a significant contribution in the EU decarbonisation path [10] and will be very useful to the achievement of the EU objectives for 2030. In order for the policy makers to develop different policies that aim at the decarbonisation goals, the Industry 4.0 tool might be an effective ally, even if not designed on-purpose to address Energy Efficiency such as other incentive mechanisms (e.g. White Certificates, ...) specifically designed to support energy efficiency in the industrial sector.

ACKNOWLEDGEMENTS

This work was financed by the Research Fund for the Italian Electrical System in compliance with the Decree of Minister of Economic Development on 16 April 2018.

REFERENCES

[1] Bazzocchi, F., Borgarello, M., Gobbi, E., Maggiore, S., Realini, A. & Zagano, C., *L'industria efficiente: Le opportunità delle imprese nella transizione energetica*, editrice Alkes, ISBN 978-88-943145-2-6, 2020.

[2] Zheng, T., Ardolino, M., Bacchetti, A., Perona, M. & Zanardini, M., *The impacts of Industry 4.0: a descriptive survey in the Italian manufacturing sector.* Journal of Manufacturing Technology Management, ISSN: 1741-038X, Dec. 2019.

[3] Zhong, R.Y., Xu, X., Klotz, E. & Newman, S.T., *Intelligent Manufacturing in the Context of Industry 4.0: A Review.* Engineering, **3(5)**, pp. 616–630, 2017. https://doi.org/10.1016/j.eng.2017.05.015

[4] Schenkel, R., *Industrie 4.0 Opportunity Discovery Workshop*; Online, https://atos.net/wp-content/uploads/2017/04/atos-industry-4-0-opportunity-discovery-workshop-brochure.pdf. Accessed on: 21 Jun 2021.

[5] Schröder, C., *The Challenges of Industry 4.0 for Small and Medium-sized Enterprises*, Division for Economic and Social Policy, 2016.

[6] Energy Strategy Group, Digital Energy Efficiency Report 2020; Online, https://www.energystrategy.it/osservatorio-di-ricerca/digital-energy-efficiency/?2020. Accesses on: 21 Jun 2021.

[7] Nota dal Centro Studi Confindustria; Centro Studi Confindustria, Numero 5/20 - 12 agosto 2020; Online, https://www.confindustria.it/wcm/connect/5a7c8fdc-2a0d-4d5a-8f9f-9b59e9824a86/Nota+CSC_Industria_4.0_120820_Confindustria.pdf?MOD=AJPERES&CACHEID=ROOTWORKSPACE-5a7c8fdc-2a0d-4d5a-8f9f-9b59e9824a86-nfxB5Ad. Accessed on: 25 Mar. 2021.

[8] Faiella, I. & Mistretta, A., *Spesa energetica e competitività delle imprese italiane*, Banca d'Italia, Numero 214, p. 22, Mar. 2014.

[9] Zagano, C., *Sistemi di gestione dell'energia. Impresa 4.0 ed efficienza energetica: a che punto siamo?*, Ricerca di Sistema, RSE n° 20000096, Milano, 2019.

[10] FINAL REPORT of the High-Level Panel of the European Decarbonisation Pathways Initiative; Online, https://ec.europa.eu/info/sites/default/files/research_and_innovation/research_by_area/documents/ec_rtd_decarbonisation-report_112018.pdf. Accessed on: 21 Jun. 2021.

A COMPARATIVE STUDY OF CONCEPTUAL DESIGN AND PROTOTYPE FOR DC-TRAD USING EV POWERTRAIN FOR RTW DC IN KT CITY

S.K. ARUN[1], I.N. ANIDA[1], P. WALKER[4], J.S. NORBAKYAH[1,2,3] & A.R. SALISA[1,2,3*]
[1] Faculty of Ocean Engineering Technology and Informatics, Universiti Malaysia Terengganu, Malaysia.
[2] Energy Storage Research Group (ESRG), Universiti Malaysia Terengganu, Malaysia.
[3] Renewable Energy and Power Research Interest Group (REPRIG), Universiti Malaysia Terengganu, Malaysia.
[4] School of Mechanical and Mechatronic Engineering, Faculty of Engineering and Information Technology,
University of Technology Sydney, Australia.

ABSTRACT

This paper is an overview of electric vehicle (EV) conceptual model development in SIMULINK; this involves components of an EV which include driver input, motor and controller, battery and the calculation of parameters with a dashboard viewing interface in which all parameters can be monitored from the dashboard. The paper focuses on comparisons of specifications and costing of an EV and a fuel-powered vehicle on route-to-work driving cycle for Kuala Terengganu city (RTW DC for KT city). A few parameters of EV were chosen to be interpreted: time, distance travelled, average speed, average running speed, average acceleration, average deceleration, acceleration percentage, deceleration percentage, idling percentage, cruising percentage, kWh and fuel costing, battery voltage, current, state-of-charge (SOC) and power. Through this, detailed overview of EV efficiency can be concluded and proven. This paper applies four methods: parameter calculation, EV modelling, data collection on RTW DC for KT city using driving cycle tracking device and validation of EV with RTW DC for KT city. The validation of the model is successful, and the travelling with the EV is proven to be more cost-efficient compared to that with fuel-powered vehicles.
Keywords: driving cycle, electric vehicle, energy, route-to-work for Kuala Terengganu, SIMULINK.

1 INTRODUCTION

Global warming and environmental deterioration are important issues faced globally. The major factor contributing to these issues is the increase in the number of vehicles on the road; hence, emissions of hazardous gasses and substances are being considered matters of concern globally. Several investigations are being conducted to study the emission rate of vehicles; the efficiency of internal burning and combustion of heavy load vehicles are found to be in satisfactory condition; however, the latter are mostly operated at lower road speeds and this reduces overall efficiency [1]. When hazardous substance emissions and engine efficiency become crucial, the invention of and research on electric vehicles (EVs) plays an important role since, using of the latter, the pollutants can be reduced drastically. In research to assess air pollution and health impacts in Malaysia, the air pollution is mainly contributed by land transportation, industrial emissions and open-burning, and of these, 70%–75% of pollution is caused by motor vehicle emissions [2]. Moreover, the most significant cause of climate change is due to excessive carbon emissions and its detrimental effects to human health [3]. The recent developments in this area were the cause for the establishment of the driving cycle. However, Malaysia is in the initial stage of developing a fuel economy driving cycle [4–6]. The discrepancies between existing driving cycles and real-world driving behaviour are caused by a few factors, such as insufficient information, improper modelling design and inaccurate driving cycle development methodologies [7]. Many models have been developed to estimate vehicle emissions and fuel consumption of a vehicle. These models can be divided into two categories, namely travel-based models or fuel-based models [8]. As a step forward for validation, driving cycle tracking device (DC-TRAD) is used to collect data on route-to-work driving cycle for Kuala Terengganu city (RTW DC for KT city) and simulated in the EV model for a comparative study.

1.1 Electric vehicle

EV runs partially or fully on electricity as a replacement and an alternative for gasoline and fuel-powered vehicles. An EV is in high demand as it consists of a smaller number of moving parts for which the maintenance cost can be reduced drastically. EVs are also environment friendly as they consume less or, in cases, no fuels and gasoline at all. Modern EVs use lithium-ion battery as it has greater reliability and excellent energy retaining capability with a minimal self-discharging rate of approximately 5% on monthly basis. Battery and electrically powered vehicles are proven to have 99% fewer moving parts, which means that they require less maintenance [9]. EVs are said to be an effective solution to reduce greenhouse effects and environment disasters [10]. Transportation and energy sectors are approximately 98% dependent on fuels and gasoline. According to United States Environmental Protection Agency (EPA), greenhouse gases (GHGs) from human activities are the most significant cause observed for climate changes since the mid-20th century [11]. To overcome this, EVs are taking over this role effectively whereby fuel-powered vehicles can be replaced by EV as this reduces emissions of GHGs by 30%–50%. Moreover, data validation is an important tool to ensure that the outcome of every research is reliable and accurate so that it can be used at all times in a proactive way [12]. As a step forward, the conceptual model of an EV is here validated and verified with the established RTW DC for KT city [13].

2 METHODOLOGY

2.1 Parameter calculation

The first step towards modelling an EV is identifying the vehicle dynamics [14]. Vehicle dynamics plays an important role in energy optimization of an EV so that it meets its main purpose in terms of energy efficiency; this involves aerodynamic forces, gravitational force, acceleration forces, rolling resistance forces, tractive effect, motor power and battery.

Aerodynamics is defined as the flow of air around and inside objects. At slow speeds, the air flowing over the vehicle body may affect the acceleration, top speed, steering control and fuel efficiency. The general aerodynamic drag force is given, in N, by

$$F_{ad} = \frac{1}{2} \rho A_f C_d V \left(S \right)^2 \tag{1}$$

where
- ρ = *air density* (kg/m^3)
- A_f = *frontal area* (m^2)
- C_d = *air drag coefficient* (dimensionless)
- $V(S)$ = *vehicle speed* (m/s)

Rolling resistance is defined in the context of energy lost for a unit distance travelled by the vehicle tire and is also known as the friction or drag force acting on the tires of the vehicle whose motion is restricted due to non-elastic effects and properties of tires. The general rolling resistance force is given by

$$F_{rr} = \mu_{rr}\, mg,$$ (2)

where
- μ_{rr} = *rolling resistance coefficient* (dimensionless)
- m = *mass of vehicle* (kg)
- g = *gravitional acceleration* (m/s^2)

The gravitational force that may act on the vehicle is given by

$$F_g = mg\, \sin\alpha$$ (3)

where
- α = grade angle with respect to horizon (°)

Besides driving resistance which occurs in steady-state motion, resistance also occurs during acceleration and braking. The factors which affect the acceleration resistance are the total mass of the vehicle and the inertial mass of rotating parts [15]. The acceleration resistance is given by

$$F_a = \left(m + \frac{\sum I_{rot}}{r^2_{dyn}} \right) \frac{dV(S)}{dt},$$ (4)

where
- I_{rot} = *Rotational components inertia* (kg/m^2)
- r_{dyn} = *dynamic radius of tires* (m)

The total tractive force is simply the summation of all forces acting on the vehicle body and is thus given by

$$F_{te} = F_{ad} + F_{rr} + F_g + F_a$$ (5)

The EV motor converts electrical power into mechanical power. The motor in an EV undergoes two phases: one during driving and one during braking. When the EV accelerates, the electric power which flows into the motor is greater than the output of the motor mechanical power, whereas during deceleration, the electric power into the motor is less than the output of the motor mechanical power. Hence, motor power output and power into motor during acceleration and deceleration are specified through the following equations:

Motor power output during acceleration,

$$P_{motorout} = \frac{P_{te}}{n_g}.$$ (6)

Motor power output during deceleration,

$$P_{motorout} = P_{te}\, n_g.$$ (7)

Power into motor during acceleration,

$$P_{motorin} = \frac{P_{motorout}}{n_m}.$$ (8)

Power into motor during deceleration,

$$P_{motorin} = P_{motorout}\, n_m.$$ (9)

where

- n_m = *motor and controller efficiency* (%)

2.2 EV modelling in SIMULINK

Automobile industries drifted towards the introduction of EV. The use of EV provides an alternative platform for those industries to transport without deteriorating the environment [16]. An EV can be modelled in various ways whereby different models are interconnected for a proper working representation of the complete system. At the initial stage of development, a few parameters have to be initialized in order to set up the vehicle and driving interface, namely tires and vehicle body parameters, type of motor being used and type of battery being integrated into the system. There are a few types of motors which can be integrated into EV modelling and can be compared on the basis of usability and dependability of the units. Most EV in the market is integrated with single induction or permanent magnet motor using an automotive differential. An automotive differential is designed to drive a pair of wheels with different speeds. An induction motor is also known as asynchronous motor which is well-known for its simplicity. The rotor of the motor consists of laminated steel with short-circuited bars which are made of aluminium in the shape of squirrel cage. The magnetic field of the stator in the motor rotates at higher speed than the rotor of the motor. Frequency is induced by the slip between the rotors and stator which produces torque to drive the motor. Permanent magnet motors are efficient and require less cooling system as the count of excitation current in it is very minimal.

Regarding batteries, most EVs use lithium-ion batteries in energy storage systems. EVs do not use a single battery as mobile phones and laptops do but a stack which consists of thousands of lithium-ion cells working in series and parallel. This works in parallel since, when the vehicle is under charge, electricity is used to convert electrical energy to chemical energy, and when the vehicle is on the road, chemical energy will be converted into electrical energy and accordingly provide sufficient power for the motor of the vehicle to produce torque and rotational speed.

2.3 Data collection on RTW DC for KT city using DC-TRAD

DC-TRAD is a device invented to collect parameters required to construct a driving cycle. DC-TRAD is known for its accuracy which is up to approximately one meter radius [17]. This device is also integrated with internet of things platform which is phpMyAdmin and MySQL database in which collected data points are updated and managed in the database system instantly. The device was used to collect data on five selected RTW DC for KT city which is approved by the Ministry of Works Malaysia as the most frequently used routes by Kuala Terengganu citizens to work [18]. These routes begin from Kampung Wakaf Tembesu and end at Wisma Persekutuan Kuala Terengganu. Figures 1–5 show the selected routes.

Figure 1: Route A.

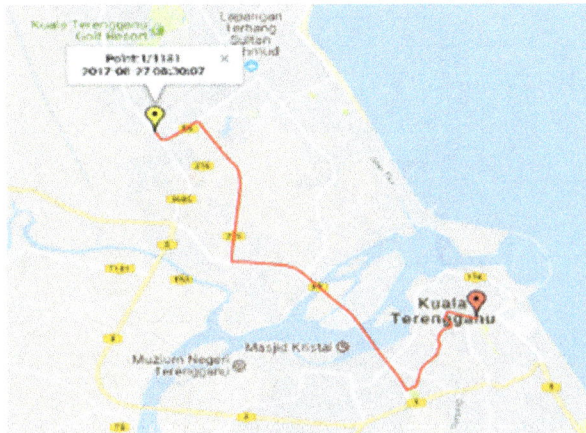

Figure 2: Route B.

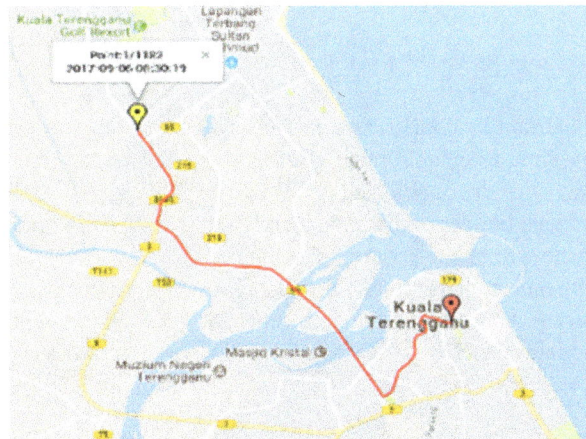

Figure 3: Route C.

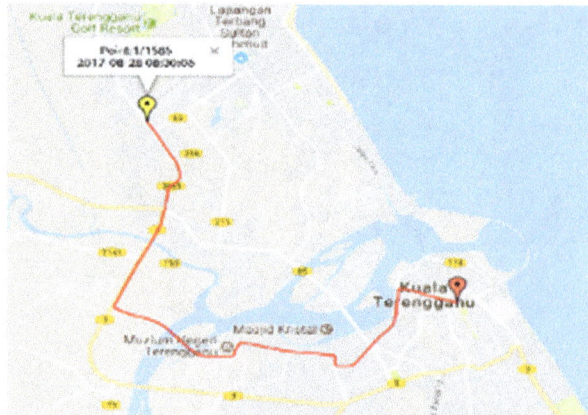

Figure 4: Route D.

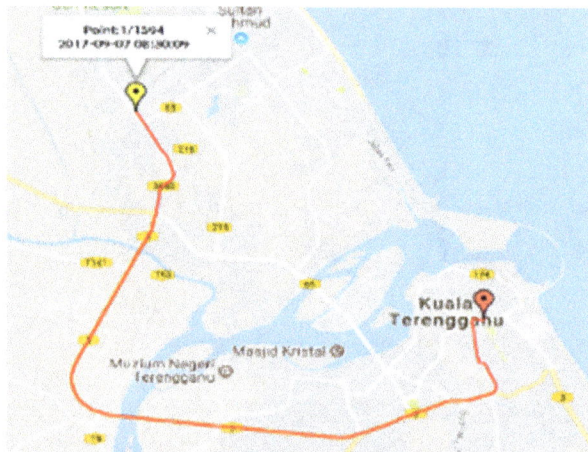

Figure 5: Route E.

2.4 Model validation with RTW DC for KT city

Data validation is important in research as it will ensure accuracy of the collected data as well as that the latter is within acceptable range. In this research, the developed EV model was verified and validated with the established RTW DC for KT city. The collected data points were uploaded into signal builder of the model, and they act as the required data, whereas the simulated driving cycle will be the acquired driving cycle.

Besides, a few simulation parameters were also chosen to be analysed; battery state-of-charge (SOC), battery parameters (voltage, current, power), kWh charges and fuel charges incurred in Ringgit Malaysia (RM) (1 RM ~ 0.24 USD) for the driving cycle route selected. As a validation procedure, percentage errors were calculated for each parameter and discussed as well.

3 RESULTS AND DISCUSSION

3.1 EV modelling in SIMULINK

Blocks from electrical simscape and vehicle dynamics blockset were used to construct a conceptual model of an EV. All required blocksets were integrated into respective subsystems to simplify the complexity of the conceptual model which comprises driver input, motor and controller, battery and vehicle body and tires. A graphical interface which acts as the dashboard of the EV with speedometer and parameter displays was arranged in a subsystem for easy viewing experience upon model simulation. Figure 6 shows the overall conceptual design of an EV.

In the driver input subsystem, a selector switch was integrated to choose between RTW DC for KT city. A drive cycle source block from Powertrain blockset is also integrated in the switch as most of the established driving cycles can be obtained from the block for verification purposes. A longitudinal driver blockset is used as the parametric speed tracking controller which generates normalized acceleration and deceleration forces according to driving cycle raw data and velocity feedbacks. Figure 7 shows the driver input subsystem.

In motor and controller subsystem, a brushed direct current (DC) motor with permanent magnet pole was used as the electric engine of the vehicle. Permanent magnet was used in an EV rotor where the current applied to the stator rotates the rotor of the motor. Brushed and brushless permanent magnet motors are commonly used in EV applications since they are known for their efficiency in terms of operation and cost. The motor is driven with a pulse width modulation (PWM) controller [19]. Square sinusoidal PWM is used in this research to create a pulse modulation of zero and one which works with the command of reference velocity; one for acceleration and zero for deceleration [20]. Figure 8 shows the motor and controller subsystem. A solver configuration block was also included in the model

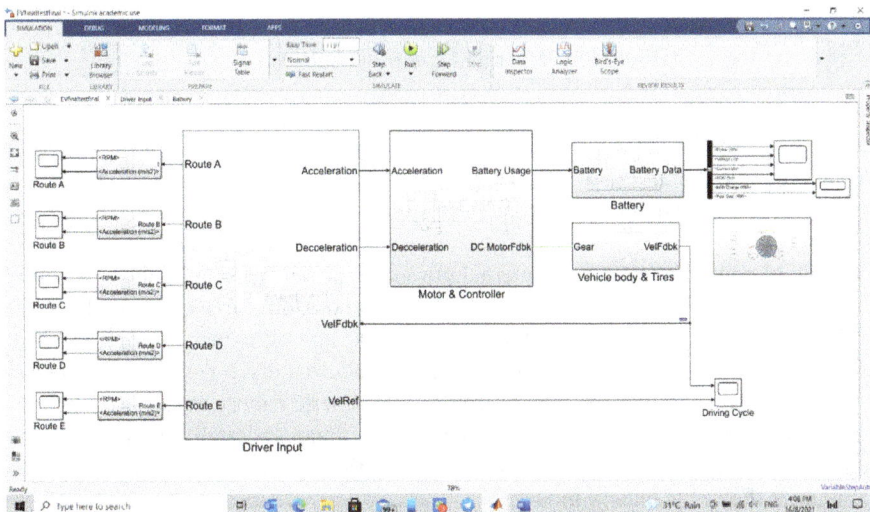

Figure 6: Conceptual model of EV in SIMULINK.

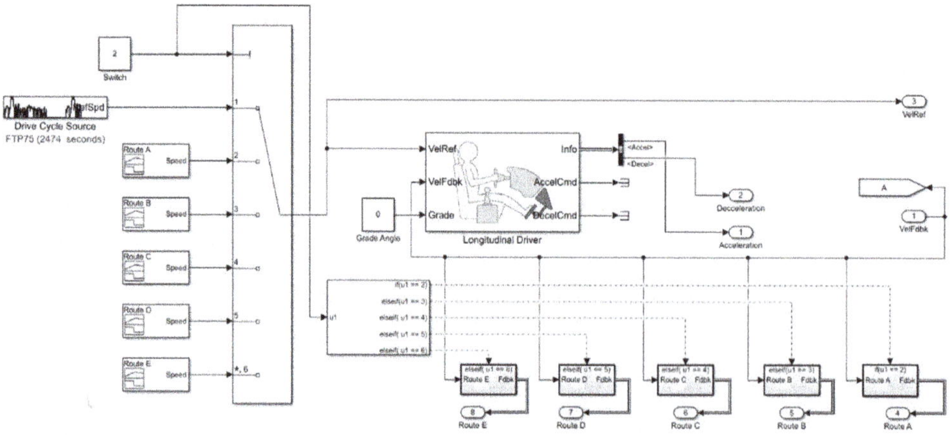

Figure 7: Driver input subsystem.

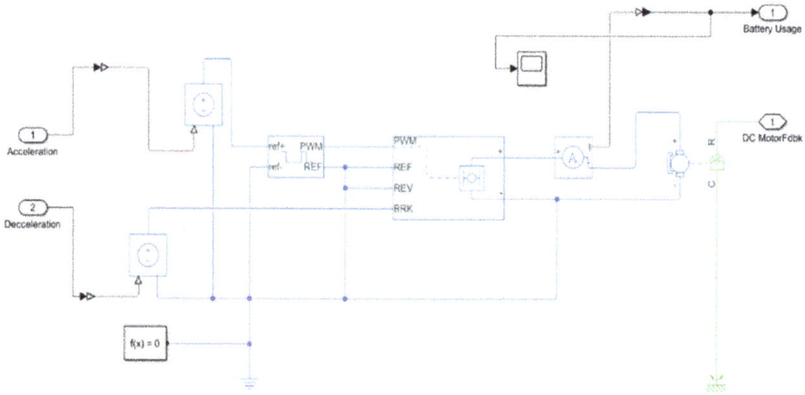

Figure 8: Motor and controller subsystem.

with ode23t solver. The solver configuration block ensures and specifies the solver parameters to be configured and that is needed to execute the model properly with minimal zero hitting and crossing and also to ensure there are not any errors encountered during the simulation.

In the PWM controller circuit, an internal power source was selected to be used with 300 VDC of power, 12×10^{-9} Ω total bridge resistance and a 0.05 Ω freewheeling diode resistance. To calculate the power consumption, battery SOC and kWh charges, a generic battery model was implemented which configures most popular battery types such as lithium-ion cells with automatic nominal parameter settings. The generic battery model was also configured in a similar way as a DC motor with 300 VDC voltage and with 100 Ah rated battery capacity. The initial battery SOC was configured at 100% with 30 seconds of battery response time. Figure 9 shows the battery model.

The total kW consumed was calculated by means of integration of power, and a typical car of Malaysia (Proton Saga) travels 9 km/L in heavy traffic [21]. The price per kW of electricity in Malaysia is at the cost of RM0.408 [22]. Thus, the total price incurred for kWh for the trip

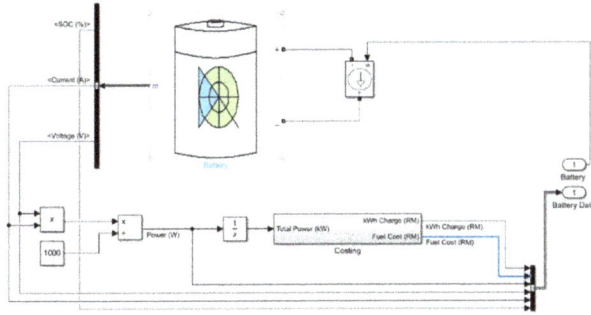

Figure 9: Battery subsystem.

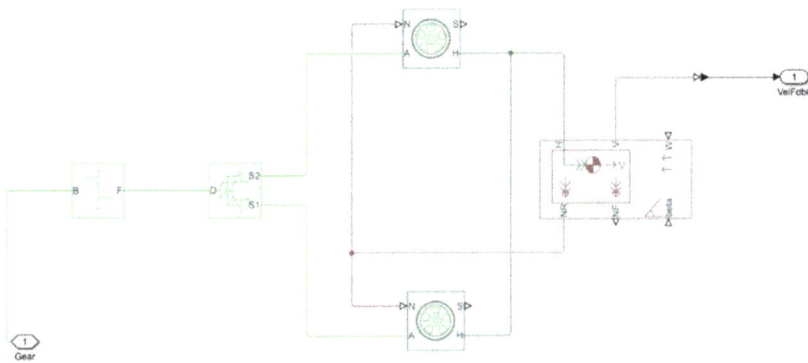

Figure 10: Vehicle and tire subsystem.

for each route was calculated as per eqn (10). The current petrol price of RON95 is placed at RM2.05/L [23]. With this, the fuel consumption price was calculated as per eqn (11).

$$kWh\ Price = kWh * 0.408 \tag{10}$$

$$Fuel\ Price = \frac{Total\ Distance\ (km)}{9} * 2.05 \tag{11}$$

A vehicle body block from simscape blockset was used, which consists of a two-axle vehicle body with longitudinal motion. Parameters such as body mass, aerodynamic drag, road grade angle and weight distribution between axles were configured in the block. Since a rear drive vehicle was implemented, two units of wheels were configured to the NR port of the vehicle body. A differential block was used with the aid of a simple gear to drive the axle of the tires. The differential block is commonly used as an additional bevel gear transmission between driveshaft and the carrier. Figure 10 shows the vehicle and tire subsystem. In addition, regenerative braking plays an important role in EV as the drive is fully dependent on electricity. Therefore, small charges and currents generated and released by the transmission of the rotor and stator of the motor is stored in the battery to be used and consumed by small power accessories such as lamps and audio systems of the vehicle.

3.2 Model validation with RTW DC for KT city

Results of required and acquired data points are as per Tables 1, 2, 3, 4 and 5.

Table 1: Route A parameters.

Item	Required	Acquired	Variance	Error (%)
Time (s)	670.000	670.000	0.000	0.000
Distance travelled (km)	13.500	13.330	0.170	1.259
Average speed (m/s)	18.074	17.230	0.844	4.672
Average running speed (m/s)	18.981	19.234	0.253	1.333
Average acceleration (m/s²)	0.979	0.952	0.027	2.729
Average deceleration (m/s²)	1.123	1.098	0.025	2.195
Acceleration percentage (%)	46.270	45.980	0.290	0.627
Deceleration percentage (%)	40.300	41.090	0.790	1.960
Idling percentage (%)	4.630	4.780	0.150	3.240
Cruising percentage (%)	8.810	8.990	0.180	2.043

Table 2: Route B parameters.

Item	Required	Acquired	Variance	Error (%)
Time (s)	826.000	826.000	0.000	0.000
Distance travelled (km)	15.120	15.300	0.180	1.190
Average speed (m/s)	12.000	11.700	0.300	2.498
Average running speed (m/s)	12.918	12.800	0.118	0.912
Average acceleration (m/s²)	0.976	1.003	0.027	2.760
Average deceleration (m/s²)	1.051	1.198	0.147	13.987
Acceleration percentage (%)	42.250	41.900	0.350	0.828
Deceleration percentage (%)	39.350	38.600	0.750	1.906
Idling percentage (%)	6.300	6.110	0.190	3.016
Cruising percentage (%)	12.110	12.100	0.010	0.083

As per simulation of route A, B, C, D and E, 95% of the acquired data are within a range of below a 5% error compared with the required data. Five percent of acquired data, highlighted in red, is out of range; as the simulation result was measured in mm/s, this is relatively small and can be neglected. Looking in detail at each driving cycle, the data recorded which is out of range is at the point of deceleration when the required data records near zero m/s and the acquired data records slightly higher which is recorded in mm/s. This is due to the inertia force on the vehicle since inertia is defined as the tendency to remain still unless an external force is applied. Inertia thus resists changes to the speed and velocity of the vehicle [22]. In

Table 3: Route C parameters.

Item	Required	Acquired	Variance	Error (%)
Time (s)	788.000	788.000	0.000	0.000
Distance travelled (km)	8.330	8.120	0.210	2.521
Average speed (m/s)	11.958	11.772	0.186	1.553
Average running speed (m/s)	13.160	13.240	0.080	0.607
Average acceleration (m/s²)	0.830	0.798	0.032	3.866
Average deceleration (m/s²)	0.999	0.928	0.071	7.107
Acceleration percentage (%)	43.910	43.122	0.788	1.795
Deceleration percentage (%)	36.420	37.540	1.120	3.075
Idling percentage (%)	10.410	10.432	0.022	0.211
Cruising percentage (%)	9.260	9.276	0.016	0.173

Table 4: Route D parameters.

Item	Required	Acquired	Variance	Error (%)
Time (s)	737.000	737.000	0.000	0.000
Distance travelled (km)	15.010	15.200	0.190	1.266
Average speed (m/s)	17.546	17.230	0.316	1.800
Average running speed (m/s)	17.960	17.880	0.080	0.446
Average acceleration (m/s²)	0.509	0.510	0.001	0.105
Average deceleration (m/s²)	0.616	0.590	0.026	4.294
Acceleration percentage (%)	27.270	26.220	1.050	3.850
Deceleration percentage (%)	23.020	23.140	0.120	0.521
Idling percentage (%)	13.030	13.143	0.113	0.867
Cruising percentage (%)	36.500	36.430	0.070	0.192

this proposed EV model, a rotor moment of inertia of 0.1 g/cm^2 was configured. The rotor inertia is inversely proportional to deceleration whereby, as the inertia increases, the deceleration decreases and the period for accelerative force to take over the vehicle control is shortened. Considering the driving cycles of all routes and calculated parameters, the model of EV is thus verified and validated as the simulated driver can follow the reference velocity closely with minimal zero crossing and hit crossing effects without compromising the acceptable percentage error which is below 5%.

Furthermore, as a generic battery model was integrated into this model, battery voltage, current, power and SOC data were also obtained. Battery size, discharge rate and SOC play an important role in a typical vehicle. In the proposed model, a battery size of 300 VDC with a 100 Ah discharge rate was chosen as most of the EV in market come with this rated battery size. Table 6 shows the battery specification analysis tabulation for routes A, B, C, D and E.

Table 5: Route E parameters.

Item	Required	Acquired	Variance	Error (%)
Time (s)	898.000	898.000	0.000	0.000
Distance travelled (km)	13.800	13.750	0.050	0.362
Average speed (m/s)	13.166	13.430	0.264	2.007
Average running speed (m/s)	13.527	13.230	0.297	2.198
Average acceleration (m/s^2)	0.685	0.710	0.025	3.578
Average deceleration (m/s^2)	0.797	0.712	0.085	10.665
Acceleration percentage (%)	43.430	43.410	0.020	0.046
Deceleration percentage (%)	37.420	37.912	0.492	1.315
Idling percentage (%)	3.230	3.130	0.100	3.096
Cruising percentage (%)	15.920	16.100	0.180	1.131

Table 6: Battery specification analysis.

Item	Unit	Route A	Route B	Route C	Route D	Route E
Peak power	kW	240.20	308.10	297.45	213.50	302.33
Peak current	A	780.40	910.15	960.34	708.80	970.23
Peak voltage	V	367.80	370.60	361.60	364.70	364.50
Battery SOC	%	95.00	93.67	95.22	94.02	95.00
kWh charge	RM	1.72	2.26	1.73	2.31	1.98
Fuel charge	RM	3.70	4.31	3.33	4.48	4.05

The maximum charging capacity of a battery was set at 350 V with internal resistance of 0.03 Ω to maintain the performance of the battery at nominal operation. This is defined as float voltage whereby the battery will be charged above the nominal voltage to ensure backup power when it is unused and to make up for the self-discharge of the battery. Float voltage is also understood to be mostly affected by the surrounding temperature. Therefore, a float voltage of 60 V was set into a generic battery model with an overall 10% slip. There were few major spikes in motor power which last for a second for each route; this is known as an over-loading of the motor. In terms of battery SOC, routes B and D used up the maximum amount of battery charge as the distance travelled in these routes was the longest, which is 15.12 km and 15.01 km, respectively, whereas route C used up the least amount of battery in which upon simulation completion, there was 95.22% of battery SOC left with the least distance travelled which is 8.33 km. On the other hand, scope for regenerative braking can be seen in the battery SOC since there are several recharging states of battery due to braking effects. The kinetic energy which is produced during braking is stored back into the battery of the vehicle so that it can be reused for acceleration and deceleration commands and for other vehicle accessories. Battery charge which is stored in the battery due to braking is relatively small, that is, within the range of 0.2–0.7 V. Besides, the costing for each route was also calculated and analysed accordingly. Overall, the cost of travelling in all routes is relatively cheaper using an EV compared to that of fuel-powered vehicles which is approximately two times higher than that of an EV.

4 CONCLUSION

To conclude, since SIMULINK has been a great platform for simulation uses in automotive industry, it was adopted in this research to design a conceptual model of an EV by which data were collected by using DC-TRAD on RTW DC for KT city, and subsequently, the efficiency of a fuel-powered vehicle and an EV on the selected routes was analysed. The proposed conceptual model has been validated and verified since the error in all calculated parameters is within acceptable range, that is, below 5% except for battery current and power due to the absence of relative resistance to limit the amount of current flowing into the electric motor and overload protection relay for the electric motor as a safety measure. By comparing the cost of travelling for all the routes, an EV is shown to be more efficient since the respective cost of travelling is two times lower compared to the cost of fuel.

ACKNOWLEDGEMENTS

The authors express their obligation to Ministry of Education Malaysia for providing financial assistance under FRGS 2020 (59623) grant and the Faculty of Ocean Engineering Technology and Informatics, UMT for all their technical and research support for this work to be successfully completed.

REFERENCES

[1] Heywood, A., Political ideas and ideologies. *Politics*, Macmillan Education, pp. 27–55, 2013. https://doi.org/10.1007/978-1-137-27244-7_2.

[2] Afroz, R., Hassan, M.N. & Ibrahim, N.A., Review of air pollution and health impacts in Malaysia. *Environmental Research*, **92(2)**, pp. 71–77, 2003. https://doi.org/10.1016/S0013-9351(02)00059-2

[3] Achour, H. & Olabi, A., Driving cycle developments and their impacts on energy consumption of transportation. *Journal of Cleaner Production*, **112**, pp. 1778–1788, 2016. https://doi.org/10.1016/j.jclepro.2015.08.007

[4] Malaysia developing draft for fuel economy driving cycle. F&L Asia, https://www.fuelsandlubes.com/malaysia-developing-draft-for-fuel-economy-driving-cycle. Accessed on: 24 November 2020.

[5] The different driving cycles – Car Engineer: Learn Automotive Engineering from Auto Engineers. Car Engineers, https://www.car-engineer.com/the-different-driving-cycles. Accessed on: 24 November 2020.

[6] Abas, M.A., Rajoo, S. & Zainal Abidin, S.F., Development of Malaysian urban drive cycle using vehicle and engine parameters. *Transportation Research Part D: Transport and Environment*, **63**, pp. 388–403, 2018. https://doi.org/10.1016/j.trd.2018.05.015

[7] Brady, J. & O'Mahony, M., Development of a driving cycle to evaluate the energy economy of electric vehicles in urban areas. *Applied Energy*, **177**, pp. 165–178, 2016. https://doi.org/10.1016/j.apenergy.2016.05.094

[8] Galgamuwa, U., Perera, L. & Bandara, S., Developing a general methodology for driving cycle construction: comparison of various established driving cycles in the world to propose a general approach. *Journal of Transportation Technologies*, **5(4)**, pp. 191–203, 2015. https://doi.org/10.4236/jtts.2015.54018

[9] What is an EV (Electric Vehicle)? – TWI, https://www.twi-global.com/technical-knowledge/faqs/what-is-an-ev. Accessed on: 2 September 2021.

[10] Jing, W., Yan, Y., Kim, I. & Sarvi, M., Electric vehicles: a review of network modelling and future research needs. *Advances in Mechanical Engineering,* **8(16)**, 2016. https://doi.org/10.1177/1687814015627981

[11] Transportation, Air Pollution, and Climate Change, US EPA, https://www.epa.gov/transportation-air-pollution-and-climate-change. Accessed on: 24 June 2021.

[12] The Importance of Data Validation, http://www.adetiq.co.uk/the-importance-of-data-validation#:~:text=Data validation is a crucial,valuable%2C demand-generating assets. Accessed on: 2 May 2021.

[13] Anida, I., Ismail, I., Norbakyah, J.S., Atiq, W.H. & Salisa, A.R., Characterisation and development of driving cycle for work route in Kuala Terengganu. *International Journal of Automotive and Mechanical Engineering,* **14**, pp. 4508–4517, 2017.

[14] Al Halabi, M. & Al Tarabsheh, A., Modelling of electric vehicles using Matlab/SIMULINK. *SAE Technical Papers,* pp. 1–10, 2020. https://doi:10.4271/2020-01-5086.

[15] Module 2: Dynamics of Electric and Hybrid vehicles Lecture 3: Motion and dynamic equations for vehicles Motion and dynamic equations for vehicles. (n.d.).

[16] Awasthi, N., Designing of electric vehicle using MATLAB and SIMULINK. Proceedings of the International Conference on Recent Advances in Computational Techniques (IC-RACT) 2020, pp. 1–10, 2020.

[17] Arun, A.K., Anida, I.N., Mariam, W.M.W., Norbakyah, J.S. & Salisa, A.R., Driving Cycle Tracking Device (DC-TRAD). *Journal of Engineering Science and Technology,* **16(4)**, pp. 2918–2926, 2021. https://doi.org/10.13140/RG.2.2.17751.60324

[18] Ministry of Works Malaysia, "2014 Road Traffic Volume Malaysia (RTVM)," Ministry of works Malaysia, Highway planning Division, 2015.

[19] Zarma, T.A., Galadima, A.A. & Aminu, M.A., Review of motors for electrical vehicles. *Journal of Scientific Research and Reports,* pp. 1–6, 2019. https://doi.org/10.9734/jsrr/2019/v24i630170

[20] Jalnekar, R.M. & Jog, K., Pulse-width-modulation techniques: A review. *IETE Journal of Research,* **3(46)**, 2000. https://doi.org/10.1080 /03772063.2000.11416153

[21] ALL CAR / VEHICLE FUEL CONSUMPTION – KADAR PENGGUNAAN MINYAK KM/LITRE, https://kereta.info/all-car-vehicle-fuel-consumption-kadar-penggunaan-minyak-kmlitre/. Accessed on: 12 February 2021.

[22] Malaysia electricity prices, GlobalPetrolPrices.com, https://www.globalpetrolprices.com/Malaysia/electricity_prices/. Accessed on: 1 December 2020

[23] Petrol Price Malaysia Live Updates (RON95, RON97 & Diesel), https://ringgitplus.com/en/blog/petrol-credit-card/petrol-price-malaysia-live-updates-ron95-ron97-diesel.html. Accessed on: 10 June 2021.

[24] Doniselli, C., Gobbi, M. & Mastinu, G., Measuring the inertia tensor of vehicles. *Vehicle System Dynamics,* **37**, 2003. https://doi.org/10.1080/00423114.2002.11666241

STRATEGIC INTELLIGENCE OF AN ORGANIZATION AMID UNCERTAINTY

LAZAR D. GITELMAN, MIKHAIL V. KOZHEVNIKOV & GALINA S. CHEBOTAREVA
Academic Department of Energy and Industrial Enterprises Management Systems,
Ural Federal University, Russia.

ABSTRACT

The paper deals with the formation and development of strategic intelligence, a fundamentally new management mechanism in organizations that provides information and analytical support for making anticipatory decisions and the company's preparedness for unpredictable challenges of the future. The paper systematizes academic approaches in terms of distinctive features and classification criteria of strategic intelligence, formulates its key objectives in the course of digital transformation, and gives the criteria for assessing its level in companies. It is shown that the establishment of strategic intelligence requires the introduction of specialized management systems, such as anticipatory management, and the formation of relevant competencies based on anticipatory learning. An anticipatory management model is proposed that takes into account weak signals for timely and adequate response to emerging threats. The power engineering industry has been used as an example for demonstrating the given model's capabilities to create standard algorithms for making anticipatory decisions in difficult situations. The paper also defines the role of strategic intelligence in the process of digital transformation and the transformation of organizations into self-learning ones.
Keywords: digital transformation, proactive management, proactive training, self-learning organization, strategic intelligence, uncertainty, weak signals.

1 INTRODUCTION

Given the context of rapid scientific and technical progress with the assumed transition to technological singularity, only those market participants who will be able to identify the upcoming changes sooner than their competitors, and to define the future shape of their industry and organization, can secure sustainable development [1–5]. Uncertainty accompanying the vision of the future forms the probabilistic nature of the intended trajectory of development and various risk constraints [6–7]. At the same time, ensuring preparedness for anticipatory response to the necessary changes and implementing them before the competitors does become not just the most important advantage but the essential condition for the survival of an organization or a company [8–10]. Such preparedness is achieved by means of using the managerial and industrial design that is flexible and adaptable to new circumstances, and the relevant unique competencies [8–10].

The said imperative is implemented when a fundamentally different development management model is introduced. This model is characterized by the focus on innovations and involves the following [3, 8, 10]:

- Regular monitoring and foresight of changes in the global environment.
- Analytical system of early identification of challenges, threats, and new opportunities as per weak signals.
- Strategic process filled with other content and methods of management.

The vision of the future with the specification required for practical purposes and the preparation for its implementation require a huge amount of new knowledge, a variety of information, and innovative solutions. Past experience and intuition-based thereon lose their significant role in management as they form a basis for traditional governance acting as a response to the deviation that has already happened and unable to prevent negative situations. The intellectual capacity of the administrative solutions will have to be radically improved within a very short time, and new management staff, systems, methods, and technologies will be needed for this purpose. However, the main thing is the business environment, in which knowledge, abilities to generate ideas and to use analytical data, and business initiative will become valuable.

2 THEORETICAL BACKGROUND, BASIC CONCEPTS, AND TERMINOLOGY

Strategic intelligence is a systematic and continuous process of exploring trends and the market environment with the use of powerful analytical systems that ensure the generation of knowledge and digital tools for making long-term decisions and the organization's preparedness for the unpredictable challenges of the future. With regard to management, strategic intelligence performs the function of detecting, identifying, and solving problems that go beyond the past experience, and which the organization is going to face in the future.

Strategic intelligence implies a coordinated combination of the following procedures:

1. Research as a tool for early response to threats and opportunities of the external environment (taking into account signals characterizing future threats taken in conjunction with the significant events of the present).
2. Comprehensive analysis of changes in consumer preferences.
3. Development of awareness (including the exchange of knowledge, experience, and projects) through the establishment of global communication networks.
4. Transformation of employees into active developers and users of innovations and strategic solutions.

In practice, there has been significant development of one of the most important components of strategic intelligence. This component is called business intelligence (BI) and describes a system of business analysis designed to support and to make managerial decisions. The term was coined in 1989 by Howard Dresner to describe 'concepts and methods to improve business decision-making by using fact-based support systems' [11]. BI pools together processes, technologies, and tools required to convert data into information, information into knowledge, and knowledge into facts that govern the operation of a successful business company. It comprises data repositories, business analytics methods, and knowledge management with the account of the following interests of different stakeholders: (1) report generation; (2) online analytical processing (OLAP); (3) data mining; (4) process mining; (5) corporate performance management, e.g. enterprise resource planning (ERP) and customer relationship management (CRM); (6) benchmarking; (7) text mining; (8) predictive and descriptive analytics; (9) scenario generation; and (10) data business design into convenient, interactive images.

BI software design makes it possible to compile quick reports, to analyze, and to submit the required data. The BI platform supports a wide range of management functions, including economic planning, budgeting, forecasting, real-time business monitoring, and scenarios for the future, advanced business analytics with the account of stakeholders' interests. These technologies make it possible to process large amounts of data in order to discover and to create new strategic business opportunities.

The emphasis is placed on traditional financial and economic indicators (e.g. profit and return on assets) when assessing the impact of BI on the economic efficiency of the company. However, it becomes particularly interesting to analyze the following non-financial benefits: development of innovative abilities and competencies; building effective communications within the organization; and reducing time for creating innovative solutions.

The quality of the company's relationships with the stakeholders (reputation and leadership, consumer satisfaction, staff motivation, and justification of innovative solutions) is a separate category of non-financial indicators [12].

A summary of academic publications reveals the following strategic intelligence functions which are of most importance to the activity of organizations:

- Prediction of fluctuations in the external environment, in the behavior of key stakeholders, or in consumer preferences, and of shifts in the competitive field for the proactive adaptation of an organization to changes and for improving its agility level [13–15].
- Information and analytical support for making anticipatory decisions, preparation of strategic reporting with the use of advanced visualization tools [16–17].
- Development of the general strategic culture of an organization, of its creative environment, encouraging open communications and distributed leadership [18–19].
- Adjustment of strategic planning and corporate policy in accordance with potential changes in the area of industrial regulation [15]. For example, the strategic intelligence of an electric utility can be used to synchronize its development plans with the prospects of the emergence of new priorities in the regional energy policy, in industrial programs of digital transformation, and in the national energy strategy.

It is important to distinguish between strategic intelligence, tactical intelligence, and operational intelligence. These types of the organization's intelligence have different process orientation, scale of the analyzed context, and tools (Fig. 1, Table 1). However, it is necessary to emphasize that strategic intelligence cannot exist without advanced operational and tactical intelligence [20]. A highly developed information infrastructure, which defines the analytical potential of the organization, is necessary for strategic intelligence to function. This potential is characterized by the so-called information width (implying the analytics' horizontal coverage of the internal processes – from routine processes to strategizing) and depth (implying the ability of analytical systems to react to trends, strong and weak signals in the core and adjacent markets, in the industries and inter-industry complexes, and finally in the global environment as a whole).

An important feature of strategic intelligence is the possibility to quantify the level of its development through the following: (a) 'prediction horizon' (the longer the prediction period, the higher the strategic intelligence); (b) depth and complexity of the changes; (c) extent of decision-making digitization (provision of digital data) and the use of predictive analytics while making strategic decisions; and (d) number of employees involved in interdisciplinary projects related to the introduction of digital technology.

Strategic intelligence may be used in firms and in organizations not only for them to achieve their market objectives but also for the purposes of ensuring the security of states [21]. In this case, it is based on powerful analytical systems supported by supercomputers and on the operation of multi-disciplinary teams bringing together experts from the areas of geopolitics, international economic activity, ecology, power engineering, systems engineering, defense, and others.

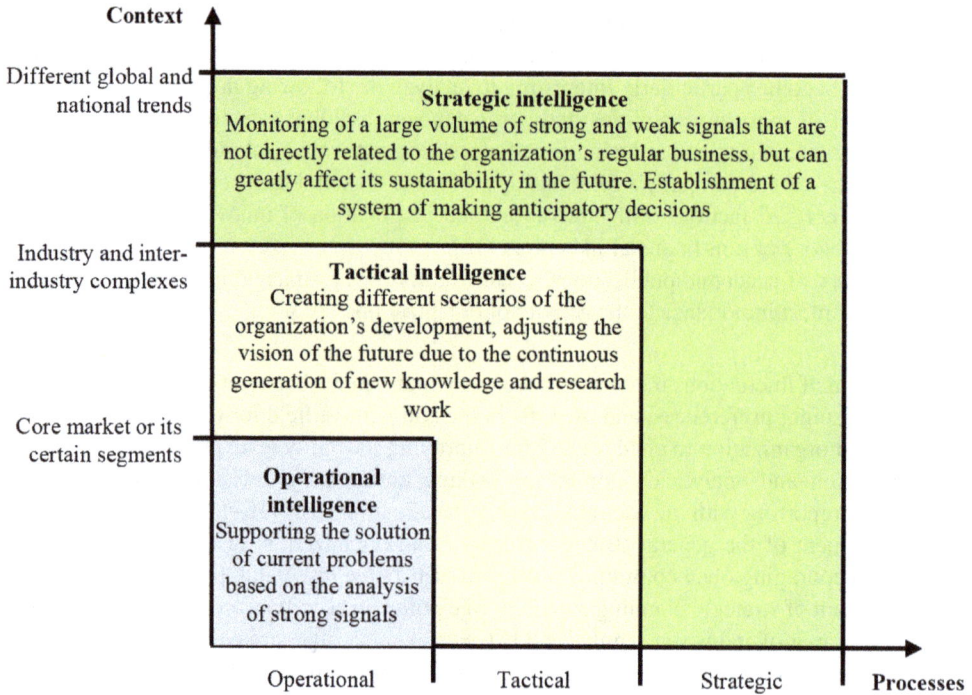

Figure 1: Focuses of different types of organization's intelligence.

Table 1: Principal differences between strategic and tactical intelligence.

Criterion	Strategic intelligence (SI)	Tactical intelligence (TI)
Main purpose	To provide the users with the information that helps to put forward new broad initiatives and to develop an outpacing strategy based on the analysis of trends within the external environment	To provide users with the information for the implementation of existing initiatives based on the analysis of the current situation
The time horizon being explored	SI provides a glimpse into the future, thus allowing organization to see emerging trends and patterns and the range of possible outcomes; it helps the organization to adjust its policy before a critical situation arises	TI works 'here and now', thus allowing organization to benefit from current advantages
Users	Persons involved in making strategic decisions; traditionally, top managers	Experts with the responsibility in a specific field of activity
Data sharpening	Moderate	High

3 INFORMATION INFRASTRUCTURE AS A BASIS FOR STRATEGIC INTELLIGENCE

Companies that already have effective strategic intelligence or wish to develop it build their own information infrastructure to be able to respond to emerging trends well in advance and to be ready to use new opportunities. Their management should assimilate various types and sources of business information, such as market, political, technological, environmental, and social, in order to visualize the future. The efficiency of how the company handles information depends on three key capabilities:

- Development of information processes for clear identification of information that is deemed strategic.
- Introduction of technologies which ensure effective information logistics.
- Building a corporate culture which encourages the exchange of information and knowledge.

However, in order to be ready to respond rapidly to unpredictable changes in the future, managers also need to develop their own research competencies, as decision-making requires a solid basis of continuously updated analytical materials and strategic foresights.

Michael Maccoby [22] proposed the following system of abilities which would enable the management aspiring to leadership to form strategic intelligence:

- Foresight – the ability to anticipate trends that can pose a threat to an organization or provide opportunities.
- Visioning – the ability to conceptualize an ideal future state and to involve others in its implementation.
- Systems thinking – the ability to perceive, synthesize, and integrate elements that function as a whole in order to achieve a common goal.
- Motivating – the ability to motivate different people to work together to implement the vision that has been created.
- Partnering – the ability to create strategic alliances with individuals, groups of people, and organizations.

The existence of blind spots – areas in which managers do not notice important information or do not understand it – is the biggest risk for strategic intelligence. These areas include, for instance, incorrect definition of the industry's boundaries, inability to identify emerging market preferences, and underestimation of competitors [23]. However, there are many ways to reduce the risks arising from the existence of blind spots. For example, it is recommended to answer the following questions for that purpose.

What are the current trends? The top managers should make sure that they understand global changes in the areas of demographics, state regulation, and consumer markets projected for the next five or ten years.

Where does the majority of industry innovations come from? How will the forthcoming developments affect the company and its business segment? The top managers should look at the adjacent markets beyond their industry, as this is where the competition will emerge in the future.

What modern technologies may lead to new forms of much-needed products or services? The same basic technologies may be used in completely different ways while developing a new product and increasing its added value.

Table 2: Organizations with high and low information infrastructure maturity [25].

Criteria	Maturity level – high	Maturity level – low
Predominant role of IT technologies	Strategic	Operational
Corporate culture	Proactivity, trust, openness to innovations are encouraged	Reactive decision-making, skepticism, resistance to changes
Defining management attention	Openness to new ideas	Control of processes and operations
Attitude toward training	Consistency of training, mistakes being structured to create the 'experience base'	Training is only necessary to know how to properly perform functions; analysis of mistakes being a waste of time
Information flows	Possess high speed and are independent on the organizational structure. Information is provided on time in the required quantity and quality	Greatly controlled and protected. There is always a lack of information, and it is provided with a delay

What are the needs of the contemporary customers? Which of these needs are satisfied, and which are not? In this regard, the analysts should constantly study the 'area', to understand the consumers' behavior and to develop the scenarios of product expansion.

What are the company's basic abilities and how can they be effectively used? It is necessary to always search for areas of further business expansion and diversification [24].

Only organizations with a high level of information infrastructure maturity, as indicated in Table 2, can carry out continuous development of strategic intelligence. Thus, the information infrastructure is a core mechanism for strategic intelligence support.

4 FOCUS ON ANTICIPATORY MANAGEMENT

Strategic intelligence describes the organization's ability to foresee the future and to be proactive. Its role increases dramatically during large-scale reforms aimed at improving the organization's sustainability in the future: e.g. it is the strategic intelligence that comes to the fore and becomes one of the core competencies during the digital transformation. In turn, the establishment of strategic intelligence requires the introduction of specialized management systems such as anticipatory management and readiness to master the relevant competencies. Figure 2 presents the key elements of anticipatory management; Fig. 3 shows how strategic intelligence propped by the anticipatory management system is involved in addressing the tasks of digital transformation.

The industry level of the anticipatory management model translates into a state policy (strategy) which defines the scientific and technical development of the given industry and its markets, mechanisms for managing reliability, safety and environmental friendliness of production, methods for regulation of mergers and acquisitions, and prices of products (for monopolists). What makes this policy different is its prominent innovative nature and its

Figure 2: Anticipatory management model.

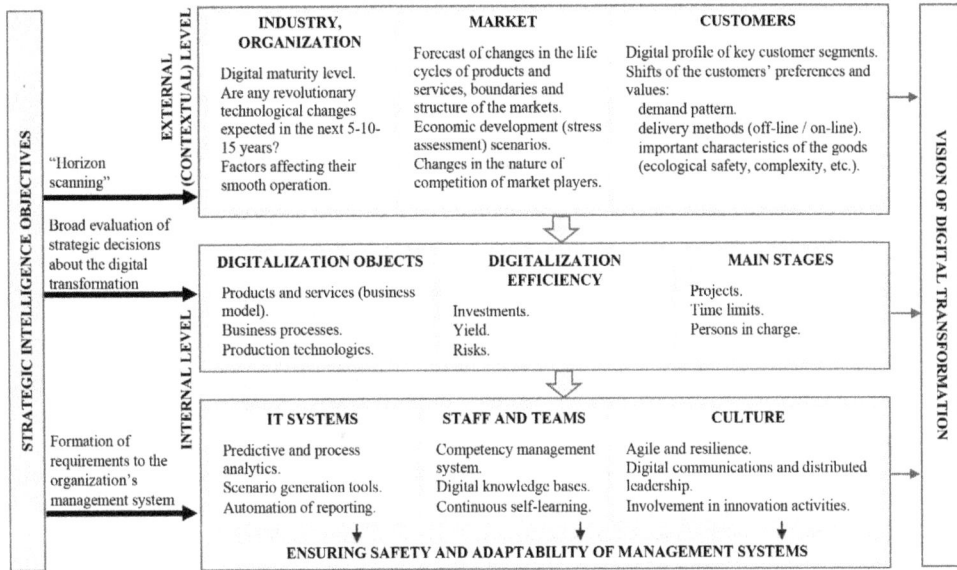

Figure 3: Strategic intelligence objectives during digital transformation.

capability to produce probability estimates and to take into account possible threats and limitations.

At the corporate level and the level of production facility, the model produces a strategy for comprehensive support and building up functional efficiency on all aspects of operation, the dependence on external factors being minimized. Particular attention should be paid to the quality and effectiveness of using resources of all kinds: equipment, fuel and power, finances and investments, and human resources. This work should be done regularly based on the continuous monitoring of the complex state of the facilities and those achievements in the scientific and technological progress which should be initiated to update production.

Thus, the concept of anticipatory management is the methodological basis for the development management model; in addition, it should provide for the sustainability of development and at the same time to increase the strategic flexibility of the organization.

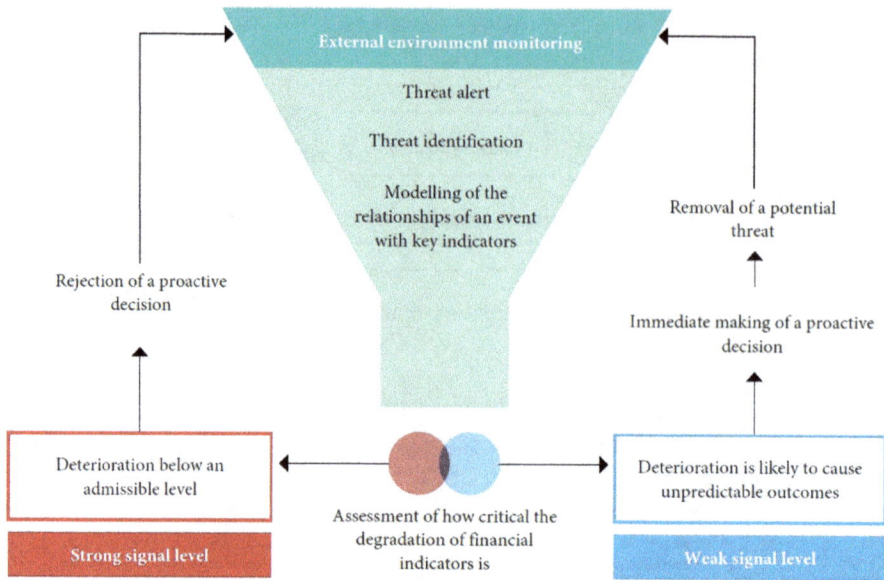

Figure 4: Anticipatory management process structure.

4.1 Proactive management model with the account of weak signals

Solving the management issue of weak signals requires a well-established surveillance system sensitive to warning information (Fig. 4); managers and staff should be prepared to embrace changes and take risk in solving new problems [26]. The problem of understanding the information contained in the signal is the most serious challenge of processing weak signals. This is due to the fact that managers' mental models are conservative and resistant to change.

Following this scheme allows one to promptly and adequately respond to emerging threats. However, monitoring is usually carried out in the strategic focus area of a company and does not preclude the risk of missing weak signals that appear in the periphery. In the power engineering industry, e.g. new technologies that become available to consumers may have a crucial impact on their businesses.

This refers to such technologies as smart power consumption metering systems, power accumulators, applications facilitating energy transactions between customers without the need to connect to market infrastructure. Accordingly, their demand for electric power decreases or increases sharply. Such changes occur everywhere, but only a few of them are actually captured by monitoring.

The application of anticipatory management technique by means of developing standard algorithms for decision-making in the power engineering industry enabled the creation of a methodological basis for anticipatory management that is shown in Table 3.

4.2 Anticipatory training that shapes strategic intelligence

Anticipatory management implies continuous anticipatory training. This term refers to an organized process of forming knowledge and competencies to solve future problems that meet global trends and national development programs. The purpose of anticipatory training

Table 3: Examples of anticipatory management algorithms.

Parameter	Wholesale generating company (TPP included)	Territorial generating company	Grid (distribution) company
Monitoring facility	Competitive positioning of the power company in the wholesale market	Balance between power and capacity (demand and supply) in the region (or electric power system)	Regulatory institution pricing policy
Threat alert	Possibility of the power company rating being decreased in terms of fuel component costs due to the emergence of a more efficient generator in the market	Expected disparities in the power balance due to the uncertainty of demand	Expected price limit for the transmission of power (capacity)
Threat identification	• Event probability estimate, % • Expected value of TPP load reduction by the market generator, MW • Event remoteness time, months	• Event probability estimate, % • Balance deficiency (redundancy) value, MW • Event remoteness time, years • Event duration – period of time prior to the increase in demand or to the launch of new power capacities, as per the original plan, years	• Probability of the decision being made by a regulatory institution, % • Relative price reduction (as compared to the current regulation method), % • Period of proactive measures (time before the restrictions are introduced), months or years • Restriction duration (estimated), years
Consequence assessment for a power company	• Net present value reduction • Reduction of future return on assets, %		
Options of proactive solutions	• Technical and organizational measures to reduce specific fuel consumption levels • Search for suppliers offering lower fuel prices • Access to the market of free contracts	• Development of demand management programs for the consumers • Emergency launch of small power producing facilities (distributed generation establishment) • Offering idle capacities to a grid operator as a system reserve • Use of temporary reduced rates for the consumers with connection delays	• Technical and organizational measures to reduce the constant part of the costs and to decrease technical and commercial losses in networks • Managing consumers' loads in order to increase the transmission capacity of the existing power grid • Contacting the consumers on the issue of reactive power (load) compensation • Building small-scale generation plants

of experts is for them to acquire knowledge for working in engineering and technical, business, and socio-humanistic systems that might be established in the foreseeable future, based on new principles and functioning in the external environment characterized by high turbulence and uncertainty [27], [28].

For this paper, it is important to emphasize that nowadays organizations are experiencing great demand for the comprehensive analytical support for business processes and solutions, which is only possible if they possess a certain level of digital maturity and high potential of strategic intelligence. The level of strategic intelligence, its update, and growth, in turn, depend on the intensity and efficiency of the research process in the organization and its partners, as well as on the timely use of new knowledge in training. It is this condition that actually ensures the relevance of the anticipatory nature of training. This is why our anticipatory training programs incorporate a distinctive research component, which is based on continuous context analysis of trends, challenges, threats, and opportunities that exist in the external environment and structural transformations happening in the economy. Such analysis is performed by students and/or staff members of companies with guidance from instructors. This exercise equips them with tools for early verification of corporate projects under development in accordance with anticipated trends.

5 CONCLUSIONS

Strategic intelligence plays a special role for the purposes of increasing sustainability and competitiveness of companies during the period of growing uncertainty of the external environment. It allows one to take into account not only the signals indicating future threats but also opportunities which will emerge in the near future. Strategic intelligence is based on digital tools for analyzing changes in consumer preferences, the transformation of employees into active developers and users of strategic solutions, and the establishment of communication networks aimed at accelerating the exchange of knowledge, experience, and projects.

Strategic intelligence is particularly important during digital transformation as it acts as a radar scanning the periphery and helps to set, most impartially, the priorities in the context of new product and investment areas, in making difficult decisions that require multi-criterion assessment of feasibility and consequences of certain actions.

The effective functioning of strategic intelligence in an organization largely depends on the availability of anticipatory training systems in it and on the ability of such systems to turn into self-learning ones. It is a company which is cultivating a continuous exchange of ideas, knowledge, and information and is continuously building up the competencies of its staff and teams, that is able to quickly adapt to market changes. Without this prerequisite, strategic intelligence would not be able to fulfil its main purpose implying foresight and readiness for the future, and risks becoming a set of costly intellectual support tools with vague functions and user-specific orientation.

ACKNOWLEDGMENT

The work was supported by Act 211 Government of the Russian Federation, contract No. 02.A03.21.0006.

REFERENCES

[1] Gitelman, L., Magaril, E., Kozhevnikov, M. & Rada, E.C., Rational behavior of an enterprise in the energy market in a circular economy. *Resources*, **8(2)**, p. 73, 2019. https://doi.org/10.3390/resources8020073

[2] Su, Y., Si, H., Chen, J. & Wu, G., Promoting the sustainable development of the recycling market of construction and demolition waste: A stakeholder game perspective. *Journal of Cleaner Production*, **277**, p. 122281, 2020. https://doi.org/10.1016/j.jclepro.2020.122281

[3] Schoneveld, G.C., Sustainable business models for inclusive growth: Towards a conceptual foundation of inclusive business. *Journal of Cleaner Production*, **277**, p. 124062, 2020. https://doi.org/10.1016/j.jclepro.2020.124062

[4] Gitelman, L.D., Silbermann, V.A., Kozhevnikov, M.V., Makarov, A.Y. & Sandler, D.G., Energy engineering and consulting: New challenges and reality. *International Journal of Energy Production and Management*, **5(3)**, pp. 272–284, 2020. https://doi.org/10.2495/eq-v5-n3-272-284

[5] Gitelman, L.D., Kozhevnikov, M.V. & Adam, L.A., Sustainable energy for smart city. *International Journal of Energy Production and Management*, **4(4)**, pp. 343–353, 2019. https://doi.org/10.2495/eq-v4-n4-343-353

[6] Chebotareva, G., Strielkowski, W. & Streimikiene, D., Risk assessment in renewable energy projects: A case of Russia. *Journal of Cleaner Production*, **269**, p. 122110, 2020. https://doi.org/10.1016/j.jclepro.2020.122110

[7] Chebotareva, G.S., Risk assessment of renewable energies: Global exposure. *International Journal of Energy Production and Management*, **4(2)**, pp. 145–157, 2019. https://doi.org/10.2495/eq-v4-n2-145-157

[8] Wang, N., Ren, S., Liu, Y., Yang, M., Wang, J. & Huisingh, D., An active preventive maintenance approach of complex equipment based on a novel product-service system operation mode. *Journal of Cleaner Production*, **277**, p. 123365, 2020. https://doi.org/10.1016/j.jclepro.2020.123365

[9] Schmitz, W.I., Schmitz, M., Canha, L.N. & Garcia, V.J., Proactive home energy storage management system to severe weather scenarios. *Applied Energy*, **279**, p. 115797, 2020. https://doi.org/10.1016/j.apenergy.2020.115797

[10] Severo, E.A., Sbardelotto, B., de Guimarães, J.C.F. & de Vasconcelos, C.R.M., Project management and innovation practices: backgrounds of the sustainable competitive advantage in Southern Brazil enterprises. *Production Planning and Control*, **31(15)**, pp. 1276–1290, 2020. https://doi.org/10.1080/09537287.2019.1702734

[11] Rouhani, S., Asgari, S. & Mirhosseini, S.V., Review study: Business intelligence concepts and approaches. *American Journal of Scientific Research*, **50**, pp. 62–75, 2012.

[12] Kuznetsov, S.Y., Business intelligence as a tool for innovative enterprise management [Biznes-intellekt kak instrument upravleniya innovacionnym predpriyatiem]. *Strategic decisions and risk management [Strategicheskie resheniya i risk-menedzhment]*, **4**, pp. 78–83, 2012.

[13] Vedder, R.G., Vanecek, M.T., Guynes, C.S. & Cappel, J.J., CEO and CIO perspectives on competitive intelligence. *Communications of the ACM*, **42(8)**, pp. 108–116, 1999. https://doi.org/10.1145/310930.310982

[14] Johanson, A., What is Competitive Intelligence, http://www.aurorawde.com/ Accessed on: 10 May 2020.

[15] Al-Zu'bi, H.A., Aspects of strategic intelligence and its role in achieving organizational agility: An empirical investigation. *International Journal of Academic Research in Business and Social Sciences*, **6(4)**, pp. 233–241, 2016. https://doi.org/10.6007/ijarbss/v6-i4/2101

[16] Baars, H. & Kemper, H.-G., Management support with structured and unstructured data - An integrated business intelligence framework. *Information Systems Management*, **25(2)**, pp. 132–148, 2008. https://doi.org/10.1080/10580530801941058

[17] Lönnqvist, A. & Pirttimäki, V., The measurement of business intelligence. *Information Systems Management*, **23(1)**, pp. 32–40, 2006. https://doi.org/10.1201/1078.10580530/45769.23.1.20061201/91770.4

[18] Rouhani, S., Asgari, S. & Mirhosseini, S.Y., Review study: Business intelligence concepts and approaches. *American Journal of Scientific Research*, **50**, pp. 62–75, 2012. https://doi.org/10.1038/scientificamerican02021884-74e

[19] Maccoby, M., Only the brainiest succeed. *Research Technology Management*, **44(5)**, pp. 1–4, 2004.

[20] Freedman, M., Creating Strategic Excellence, http://www.kepner_tregoe.com\Accessed on: 10 May 2020.

[21] Akhgar, B. & Yates, S., (eds.), *Strategic Intelligence Management: National Security Imperatives and Information and Communications Technologies*, Elsevier: Oxford, 2013.

[22] Maccoby, M., (ed.), *Strategic Intelligence, Conceptual Tools for Leading Change*, Oxford University Press, 2015.

[23] Jones, J., (ed.), *Design Methods [Metody proektirovaniya]*, Mir [Mir]: Moscow, 1986.

[24] Büchel, B., From blind spots to strategic intelligence. Ensuring growth options are exploited, https://www.imd.org/research-knowledge/articles/from-blind-spots-to-strategic-intelligence/ Accessed on: 10 May 2020.

[25] Gileva, T.A., Digital maturity of the enterprise: Methods of assessment and management. *Bulletin USPTU. Science, Education, Economy. Series Economy*, **1(27)**, pp. 38–52, 2019. https://doi.org/10.17122/2541-8904-2019-1-27-38-52

[26] Gitelman, L.D., Gavrilova, T.B., Gitelman, L.M. & Kozhevnikov, M.V., Proactive management in the power industry: Tool support. *International Journal of Sustainable Development and Planning*, **12(8)**, pp. 1359–1369, 2017. https://doi.org/10.2495/sdp-v12-n8-1359-1369

[27] Gitelman, L.D., Kozhevnikov, M.V. & Ryzhuk, O.B., Advance management education for power-engineering and industry of the future. *Sustainability*, **11(21)**, p. 5930, 2019. https://doi.org/10.3390/su11215930

[28] Senge, P., (ed.), *Fifth Discipline: The Art and Practice of Self-learning Organizations [Pyataya disciplina: iskusstvo i praktika samoobuchayushchejsya organizacii]*, Olymp-Business [Olimp-Biznes], 1999.

NEM SCHEMES ANALYSIS BASED ON INSTALLED GRID-CONNECTED PV SYSTEM FOR RESIDENTIAL SECTOR IN MALAYSIA

W. M. W. MUDA[1,2], N. ANANG[2] & AIDY M. MUSLIM[3]
[1] Renewable Energy and Power Research Interest Group (REPRIG), Eastern Corridor Renewable Energy SIG.
[2] Faculty of Ocean Engineering Technology and Informatics, Universiti Malaysia Terengganu,
21030 Kuala Nerus, Terengganu, Malaysia.
[3] Institute of Oceanography and Environment (INOS), Universiti Malaysia Terengganu,
21030 Kuala Nerus, Terengganu, Malaysia.

ABSTRACT

An investigation has been conducted to analyse the performance of a grid-connected photovoltaic system (GCPV) based on the net energy metering (NEM) scheme. Several analyses of a similar system have been performed in the literature based on assumptions and simulations. However, the concept based on actual NEM data in Malaysia has not been fully considered. Hence, this study analyses the real performance of the GCPV system from the field monitoring of PV energy production, as well as import and export energy, collected at a residential house participating in NEM 2.0. From the collected data, the economic parameters were calculated and compared with an equivalent system before the NEM implementation, which is a grid-only system, and the NEM 1.0 and 3.0 schemes. The results show that for the considered load demand with an average monthly electricity bill of RM 500, the NEM 2.0 provides more benefits to consumers with the lowest payback period, net present cost, net saving and energy cost. Although NEM 3.0 produced the lowest net saving, which was RM 33,280 for 20 years of the project's lifetime, it was still capable of reducing the electricity bill by 66% for the first year and 32% during self-consumption.

Keywords: economic analysis, grid-connected PV system, Kuala Terengganu, net energy metering.

1 INTRODUCTION

The net energy metering (NEM) concept was launched in the 1980s in the United States to reduce dependency on the conventional grid, to promote green energy technology and to encourage on-site distributed generation, especially from rooftop solar panels and small-scale renewable energy (RE) sources [1]. The main benefit of the NEM is to reduce the monthly electricity bill by consuming the energy produced by the RE system. Any shortage of electricity is backed up by importing the energy from the grid, and an excess of electricity is exported back into the grid. The net meter calculates the net energy used by the consumer at the end of the month. In most cases, residential NEM consumes more power than RE production; thus, consumers still have to pay a small amount of electricity bill [1].

Several studies have been conducted in the literature to analyse the performance of PV systems under the NEM scheme in Malaysia [2–5], especially for residential areas. Authors in [4] compared Feed-in-Tariff (FiT) and NEM schemes in terms of the payback period for low- and medium-load consumption. Since the FiT rate was assumed to decrease by 8% every year until the period ended, there were cases where the payback period of the FiT scheme was higher than that of the NEM. In reality, the FiT rate is fixed for 21 years once the commencement date has been achieved [6]. The rate decreases every year due to the decrease in RE technology market price, but this is only applicable to new Feed-in-Approval Holder (FiAH). Mansur *et al.* [5] investigated the technical, economic and environmental aspects. The considered load profile was low, and the authors varied the size of the capacity from 2 to 12 kWp. It is true that for low-load demand and high PV capacity, there is a possibility that

© 2022 WIT Press, www.witpress.com
DOI: 10.2495/EQ-V6-N4-382-394

customers do not have to pay for their electricity bill as it reaches zero. Nonetheless, the higher the PV capacity, the higher the installation costs and maintenance costs, which were not considered in their study. Thus, the longer the payback period, the longer the customers have to endure due to low savings. Razali *et al.* [2] proposed integration of time of use (TOU) into the NEM scheme to improve the original NEM. However, the highlighted problem of NEM was not accurate by comparing low- and high-load profiles using the same capacity of the PV system. In reality, the low-load profile should be installed with a low PV capacity so it will yield the same benefit as the high-load profile with a high PV capacity. Razali *et al.* [7] compared NEM 1.0 and NEM 2.0 using three different sizes of customers and PV capacity of 1–8 kWp. The period of simulation is 25 years. However, the calculated bill equations for NEM 2.0 were not correctly written. In their paper, the net power from the net meter was calculated before being multiplied with the price tariff. The actual calculation should be that the imported energy is multiplied with the gazette tariff, and the export bill is calculated by multiplying the exported energy with the gazette tariff in descending order [8]. Thus, this misconception affected the results obtained. In addition, the authors of [3] calculated the PV production from PV-rated power. Nevertheless, from the literature [9], the annual capacity factor, which is the ratio of the actual annual energy output to the amount of energy that the system would produce at full-rated power, shows that the value is low, around 20%. Hence, it is not accurate if the full-rated power is used in the calculation to find the payback period, as it is far from being true.

From the above-mentioned literature, it is obvious that all these misconceptions were due to unclear explanations on authorities' websites, such as the Sustainable Energy Development Authority (SEDA) Malaysia [6]. In addition, previous research was based on simulations and assumptions. The novelty of this research work is the performance of the PV system assessed based on real data collection from field monitoring of a residential house participating in NEM 2.0. In addition, the correct way to calculate the actual monthly electricity bill for NEM 2.0 is shown. From the collected data, the economic parameters of NEM 2.0 are calculated and compared with the grid-only system. Then, using the same data, the economic parameters for NEM 1.0 and NEM 3.0 are computed for comparison purposes.

2 NET ENERGY METERING

NEM is one of the renewable energy policies to encourage consumers to produce their own electricity. It is also known in the literature as net metering, or net FiT [10]. The main difference between net FiT and gross FiT is the amount of electricity fed into the grid. If all the electricity generated is fed into the grid and the consumers purchase any electricity to consume from the grid, then it is gross FiT. Meanwhile, in the NEM or the net FiT scheme, only the excess electricity is exported into the grid. In both schemes, the electricity producers are paid at a certain rate for any injected electricity into the grid.

The export rate is different for different countries, and it depends on many factors. In certain countries, the export rate depends on the capacity of the plant, types of RE sources and different sectors [11]. The authors of [3] and [7] have listed export rates for different countries, such as Australia, Canada, Cyprus, Greece and the Netherlands. Table 1 shows additional countries with different import and export rates.

There is also literature research on finding the best export rate to increase electricity producers from RE, such as in [15] and [16]. In Malaysia, the export rate is different based on different schemes. For the gross FiT scheme that was introduced in 2011, the export rate was so high compared to the import rate that it boosted the RE producers among consumers from 649 MW in 2012 [17] to 2,072 MW in 2017 [18]. Although the FiT rate has decreased every

Table 1: Comparison of import and export rates for different countries.

Country	Export rate	Electricity tariff
France [12]	0.05 €/kwh = 0.24 RM/kWh	0.08 €/kwh = 0.39 RM/kWh
India [11]	Rs. 15/kwh = 1.23 RM/kWh	Rs. 4/kWh = 0.33 RM/kWh
Iran [13]	0.08 $/kWh = 0.33 RM/kWh	0.05 $/kWh = 0.21 RM/kWh
Spain [14]	0	0.2477 €/kWh = 1.21 RM/kWh

Table 2: Domestic tariff rate [19].

R_i	Block tariff	Rate (RM/kWh)	Service tax	RE fund
R_1	200	0.218		
R_2	100	0.334	0%	
R_3	300	0.516		1.6%
R_4	300	0.546	6%	
R_5	> 900	0.571		

year due to the decrease in RE technologies, it has received encouraging responses from Malaysians, especially in the residential sector, which has seen the highest percentage of FiT applications at 85.85% in the SEDA Annual Report 2017 [18]. To reduce the burden of the government paying a high FiT rate, the NEM scheme has been implemented in Malaysia since 2016.

2.1 NEM 1.0

The NEM concept was introduced in Malaysia not only to reduce dependency on imported fossil fuels but also to reduce monthly electricity bills if there is any possibility of future increase in electricity tariffs. The NEM is executed by the Ministry of Energy and Natural Resources (KeTSA), regulated by the Energy Commission (EC), with the SEDA Malaysia as the implementing agency.

The government offers a 500 MW quota for the PV system under this scheme. In NEM 1.0, every exported energy unit into the grid will be credited in the next billing at a displaced cost of 0.31 RM/kWh unit.

As shown in Table 2, the electricity purchase rate in Malaysia is based on a different block. The higher the load consumption, the higher the rate. From the table, it can be seen that the 6% service tax is not applied to the first 600 kWh and 1.6% is applied to the total cost of electricity used as the RE fund collected by the government to promote the growth of electricity generation from RE.

Thus, NEM 1.0 is beneficial for low-load consumption, especially if it is lower than 200 units, as the purchase rate is lower than the sellback rate. However, for high-load consumption, a high tariff rate is applied, so it is not financially appealing.

The low performance of NEM 1.0 can be seen from the SEDA annual report 2018 [20], where only 5.6% of the 500 MW quota has been approved. In the same report, the new concept was introduced in NEM 2.0.

2.2 NEM 2.0

To overcome the disadvantages of NEM 1.0, the government of Malaysia has improvised the concept of NEM 2.0, starting on 2nd January 2019. This scheme is only applicable to Peninsular Malaysia consumers who registered with Tenaga Nasional Berhad (TNB), where the export bill is calculated using the same tariff as in Table 2, but in reverse order. Thus, for higher load consumption, a higher rate will be applied to the export bill. Due to this improvement, the quota of 500 MW was discontinued 1 month before the end date of 31st December 2020. Similar to NEM 1.0, any excess electricity can be rolled over for a maximum period of 24 months. After that period, the excess energy will be forfeited. The offset period of this scheme is also 20 years, as in the previous scheme.

2.3 NEM 3.0

Starting on 1st February 2021, the government has offered another 200 MW of capacity from solar systems under NEM 3.0. This quota is further divided into two, which are 100 MW under NEM Rakyat (residential building) and another 100 MW under NEM GoMEn (government building). Another 300 MW is reserved to be offered on 1st April 2021 for NOVA, which is for commercial and industrial sectors. This quota is open for 3 years. The biggest change in this scheme is the offset period, which is only 10 years. It means, for the first 10 years, the concept is similar to NEM 2.0, where the excess electricity is exported into the grid and will be credited on the next bill at a retail rate in reverse order.

After 10 years, the system uses the self-consumption (SELCO) concept. During SELCO, the excess electricity can be exported into the grid, but no export bill is considered. Thus, the consumer is encouraged to fully utilise PV production. The second change is that the rollover period is reduced to 12 months, and there is no rollover during the SELCO. To enhance the self-consumption of rooftop solar PV prosumers, there is a necessity to incorporate a battery energy storage system into the solar system [16].

3 METHODOLOGY

In this study, the performance of the installed grid-connected PV (GCPV) system of a house with a NEM 2.0 scheme was investigated. The specification of the PV system is summarised in Table 3.

From 2 meters provided during the installation of the system, three types of data were collected, namely the PV production (E_{PV}), energy import (E_{import}) and export (E_{export}). Daily values of these data are available to the owner of the house. The monthly data for these parameters for the year 2020 are shown in Fig. 1. From these data, PV energy self-consumed ($E_{pv,selco}$) and total energy used by the load (E_{load}) can be calculated from eqns (1) and (2), respectively.

$$E_{pv,selco} = E_{PV} - E_{export} \tag{1}$$

$$E_{load} = E_{import} - E_{pv,selco} \tag{2}$$

Total energy consumption by the load and the grid can be written as in eqn (3), and the total energy production by PV and the conventional grid is given in eqn (4).

$$Total\ consumption\ (kWh) = E_{load} + E_{export} \tag{3}$$

Table 3: Specification of installed PV system.

Parameters	Values
Total PV capacity	6.12 kWp
1 unit PV capacity	340 W
Number of PV module	18 units
PV lifetime	21 years
Roof covered area	36 m^2
Inverter capacity	1 unit 6 kW
Inverter lifetime	10–15 years
Project lifetime	20 years
Capital cost	RM 26,592.00

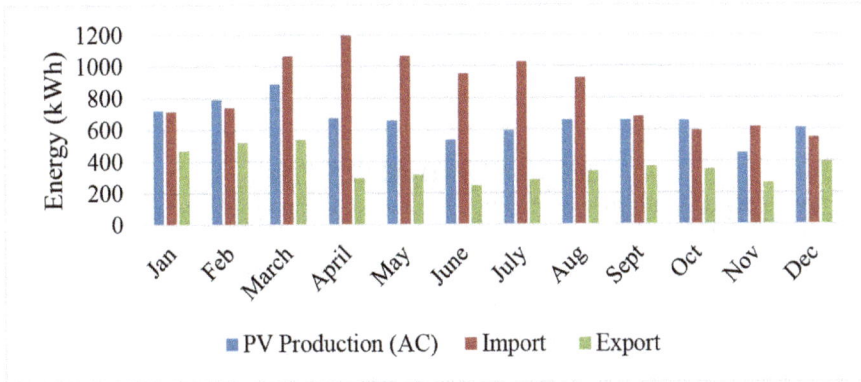

Figure 1: PV production, import and export energy collected for the year 2020.

$$Total\ production\ (kWh) = E_{PV} + E_{import}. \tag{4}$$

From E_{load}, the electricity bill before the PV system was installed ($Bill_{beforeNEM}$) can be estimated from a standard calculation using the electricity tariff as in Table 2. The monthly savings can be obtained from the collected data based on eqn (5).

$$Saving = Bill_{beforeNEM} - Bill_{afterNEM}. \tag{5}$$

Then, a simple payback period (*PB*) that shows the shortest period for the cumulative economic savings to become equal to the total initial investment can be computed [9] using the following equation:

$$PB = \frac{Capital\ cost}{saving} \tag{6}$$

The saving in eqn (5) is not a net savings, but based on monthly savings obtained from electricity bill reduction. The more accurate way to calculate the savings is by considering all costs during the project's lifetime as a net present cost (*NPC*). The *NPC* value considers all costs during the project's lifetime, including the capital cost, maintenance and operation costs (O & M), replacement costs, salvage and grid sales revenue. Neglecting the annual real discount rate, the *NPC* can be calculated using the following equation.

$$NPC\ (RM) = Total\ expenditure\ (RM) - Total\ revenue\ (RM). \qquad (7)$$

Another important parameter is the cost of energy (*COE*), which shows how much the consumer has to spend per kWh.

$$COE\left(\frac{RM}{kWh}\right) = \left(\frac{Annualized\ total\ cost\ (RM)}{Total\ consumption\ (kWh)}\right) \qquad (8)$$

where the total consumption is given in eqn (3) and annualised total cost can be obtained from the *NPC* value. The annual net savings can then be obtained from the following equation:

$$Saving_{net}(RM) = \frac{NPC_{beforeNEM} - NPC_{afterNEM}}{Project\ lifetime} \qquad (9)$$

4 RESULTS AND DISCUSSION

This section is divided into four parts. First, the economic analysis before installing the PV system is presented. Next, based on the actual NEM 2.0 electricity bill, the correct way to calculate the export bill is presented. Then, the performance of NEM 2.0 is discussed based on data collection. The comparison between all three NEMs is elaborated in the last subsection. All results were obtained by calculation based on a year's data collection. The assumption was made that all collected data would remain for 20 years.

4.1 Grid-only system

Before analysing the impact of the NEM system, first, the economic parameters using a grid-only system were analysed. For the annual energy consumption, E_{load} was calculated from eqn (2), which was 13,654 kWh. Then, the annual electricity bill before NEM had been implemented was calculated using Table 2, showing a value of RM 6,701/year. The *NPC* of this system for 20 years was obtained by multiplying the annual bill before NEM by 20 years, neglecting the initial cost incurred by the consumer during the installation of the grid connection. Assuming the bill remains for 20 years, the $NPC_{beforeNEM}$ would be RM 134,020. From the annual bill and energy consumption, the $COE_{beforeNEM}$ was obtained, showing a value of 0.4908 RM/kWh.

4.2 Export bill calculation

This section shows how the NEM 2.0 electricity bill was calculated. It depends on E_{import} and E_{export}. First, the import bill was calculated similar to a standard electricity bill using Table 2. To encourage RE producers to use the NEM scheme, the government calculates the export bill in reverse order starting with the higher block tariff, subjected to the maximum block tariff used in the import bill.

Two examples are given here based on the actual bill. From Fig. 1, the electricity bills in April and December were taken as examples, where in April, E_{import} was the highest at 1,196 kWh but E_{export} was low, while in December, E_{import} was 548 kWh and E_{export} was 398 kWh. The electricity bill before tax in April 2020 is presented in Table 4. Since the highest rate for the import bill was R_5, the export bill was calculated starting from that rate. The total exported energy in April was only 295 kWh, which was less than the amount of imported energy at R_5. Hence, the total export energy was multiplied by that rate.

In December 2020, the import bill was calculated, as shown in Table 5. Since the imported energy was low, the maximum rate was R_3. Thus, R_3 was the highest rate used in the calculation of the export bill in December. Since the E_{import} for the R_3 was only 248 kWh and the total E_{export} was higher than that, the balance of usage was multiplied with the lower rate of R_2. Since the maximum usage block of R_2 was 100 kWh, the remaining 50 kWh was multiplied with the lowest rate of R_1.

In general, the relationship between import and export bills is represented in Table 6. Fifteen different cases may occur. For example, in reference to case 1, if the import bill is high and extends up to R_5, the export bill will start at R_5 with the same number of electricity units or less, depending on the value of exported energy. If the exported energy is high, it extends the rate up to R_1.

To prevent the rollover of excess energy to the next billing statement, the exported energy must always be less than or equal to the imported energy. Consequently, NEM customers

Table 4: Calculation of import and export bill in April 2020.

	Import bill				Export bill		
R_i	Rate (RM/ kWh)	Usage (kWh)	Amount (RM)	R_i	Rate (RM/ kWh)	Usage (kWh)	Amount (RM)
R_1	0.218	200	43.60	R_5	0.571	295	168.45
R_2	0.334	100	33.40	R_4	0.546	0	0.00
R_3	0.516	300	154.80	R_3	0.516	0	0.00
R_4	0.546	300	163.80	R_2	0.334	0	0.00
R_5	0.571	296	169.02	R_1	0.218	0	0.00
Total		1,196	564.62	Total		295	168.45

Table 5: Calculation of import and export bill in December 2020.

	Import bill				Export bill		
R_i	Rate (RM/ kWh)	Usage (kWh)	Amount (RM)	R_i	Rate (RM/ kWh)	Usage (kWh)	Amount (RM)
R_1	0.218	200	43.60	R_3	0.516	248	127.97
R_2	0.334	100	33.40	R_2	0.334	100	33.40
R_3	0.516	248	127.97	R_1	0.218	50	10.90
Total		548	204.97	Total		398	172.27

Table 6: Fifteen different cases in import and export bills of NEM.

Case	Import bill					Export bill				
	R_1	R_2	R_3	R_4	R_5	R_5	R4	R_3	R_2	R_1
1	√	√	√	√	√	√	√	√	√	√
2	√	√	√	√	√	√	√	√	√	
3	√	√	√	√	√	√	√	√		
4	√	√	√	√	√	√	√			
5	√	√	√	√	√	√				
6	√	√	√	√			√	√	√	√
7	√	√	√	√			√	√	√	
8	√	√	√	√			√	√		
9	√	√	√	√		√				
10	√	√	√					√	√	√
11	√	√	√					√	√	
12	√	√	√					√		
13	√	√							√	√
14	√	√							√	
15	√									√

have to ensure that the installed capacity of the PV system is not too high to prevent exporting the excess limit. However, if that happens, the bill will be zero and the excess energy will be credited to the next billing statement and can be kept for up to 24 months for NEM 2.0. Using this concept, the average export rate is always higher than or equal to the import rate.

4.3 NEM 2.0 scheme

From the collected data that is under the NEM 2.0 scheme, the important economic parameters were analysed using this scheme before comparing it with NEM 1.0 and 3.0. From eqns (1) and (2), the load energy consumption, E_{load} from different components, $E_{pv,selco}$ and E_{import} can be seen in Fig. 2. From the figure, the PV fraction contributes to 23–34% of the total load depending on the availability of solar irradiance and electricity usage.

The electricity utility provides the import and export bills and also $Bill_{afterNEM}$. Since $Bill_{beforeNEM}$ has been calculated in the previous section, then, all the bills are depicted in Fig. 3. Referring to Figs. 2 and 3, it can be seen that the $Bill_{beforeNEM}$ and the import bill are highly dependent on the load consumption. However, the export bill depends on the availability of PV production and the load demand.

The annual data collection and computed parameters can be summarised in Table 7. From the table, the annual $Bill_{beforeNEM}$ was reduced to RM 2,270 for $Bill_{afterNEM}$, approximately a 66% reduction. This reduction can be considered as an annual savings benefit from NEM 2.0, which was calculated from eqn (5). From this savings, the PV producer normally calculates the simple payback period in eqn (6), and in this case, the value is 6 years.

However, the annual saving in Table 7 is not considered as net savings because it does not include all the costs that must be spent during the project's lifetime, especially the

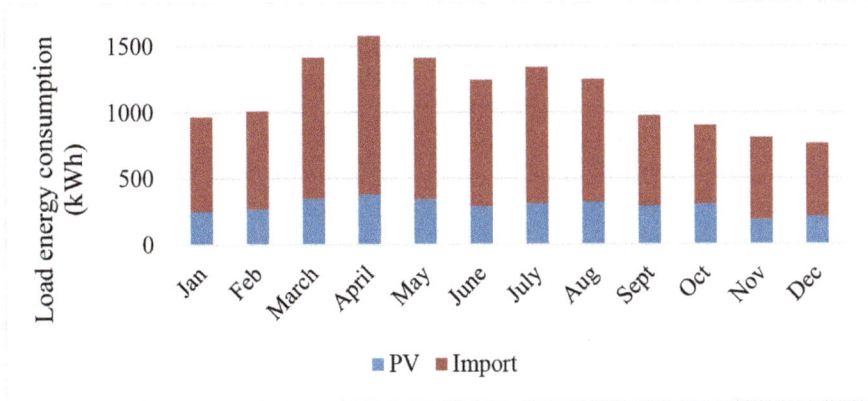

Figure 2: Fraction of energy used by the load.

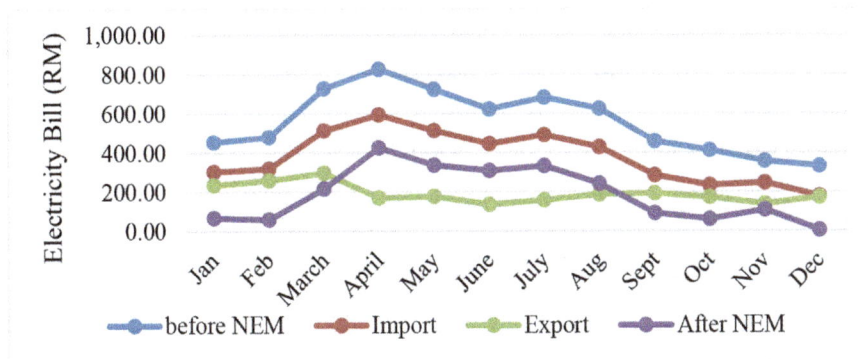

Figure 3: Electricity bill before and after NEM, import and export bills.

replacement cost of the inverter, which has a shorter lifespan. Thus, *NPC* was calculated as shown in Table 8. Assuming that the inverter is replaced after 10 years, and all the data collection is maintained every year, then the $NPC_{afterNEM}$ for 20 years can be calculated and the value would be RM 79,392. Note that the O & M cost refers to the $Bill_{afterNEM}$ for 20 years that must be spent on imported energy from the grid.

The advantage of NEM can be seen clearly from the annual net saving using eqn (9), with a value of RM 2,731. The *COE* after NEM is also reduced to RM 0.22 per kWh, which shows that the cost of electricity is cheaper by integrating the PV system and the grid.

4.4 Comparison between NEM 1.0, NEM 2.0 and 3.0

Using the same collected E_{PV}, E_{export} and E_{import}, this section compares the performance of the three NEM schemes to see the effect of the parameters when changes are applied to policies, assuming that all parameters remain the same for different schemes. The first parameter is the simple payback period. It can be seen clearly in Fig. 4 that NEM 1.0 has a higher simple

Table 7: Annual data collection from NEM 2.0 users in the year 2020.

Parameters	Values
Annual E_{PV}	7,913 kWh
Annual E_{import}	10,132 kWh
Annual E_{export}	4,391 kWh
Annual $E_{pv,selco}$	3,522 kWh
Annual E_{load}	13,654 kWh
Annual $Bill_{beforeNEM}$	RM 6,701.00
Annual import bill	RM 4,405
Annual export bill	RM 2,305
Annual $Bill_{afterNEM}$	RM 2,270
Annual saving bill	RM 4,431

Table 8: The $NPC_{afterNEM}$ components.

Component	PV	Inverter	Grid	System
Capital (RM)	19,192	7,400	0	26,592
Replacement (RM)	0	7,400	0	7,400
O & M (RM)	0	0	45,400	45,400
Salvage (RM)	0	0	0.00	0.00
Total (RM)	19,192	14,800	45,400	79,392

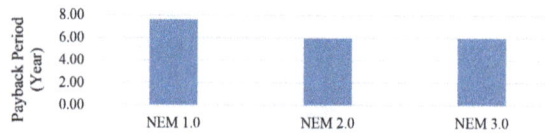

Figure 4: Simple payback period for different NEM schemes.

payback due to the lower export rate, which is fixed at RM 0.31. Hence, using the same amount of annual export energy as in Table 7, the annual export bill is RM 1,361. Therefore, the reduced annual bill for NEM 1.0 is RM 3,214. The increase in the $Bill_{afterNEM}$ reduced the annual savings from RM 3,487, thus increasing the payback period to 7.63 years. Meanwhile, for the first 10 years of NEM 3.0, everything would be the same as in NEM 2.0. Accordingly, the simple payback period is similar to that of NEM 2.0.

The NPC value, which was calculated over the project's lifetime of 20 years, was plotted in Fig. 5. For NEM 1.0, the capital and replacement costs are the same as in NEM 2.0. However, the O & M cost, which refers to the electricity bill paid by the consumers, was higher at RM 64,280 for NEM 1.0 compared to that of NEM 2.0, thus the NPC is higher. For NEM 3.0, for the first 10 years, every cost is similar to NEM 2.0. Nonetheless, for the next 10 years, the

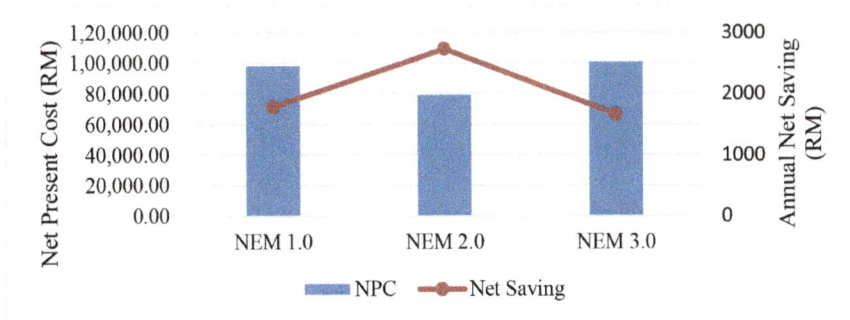

Figure 5: *NPC* and annual net saving for different NEM schemes.

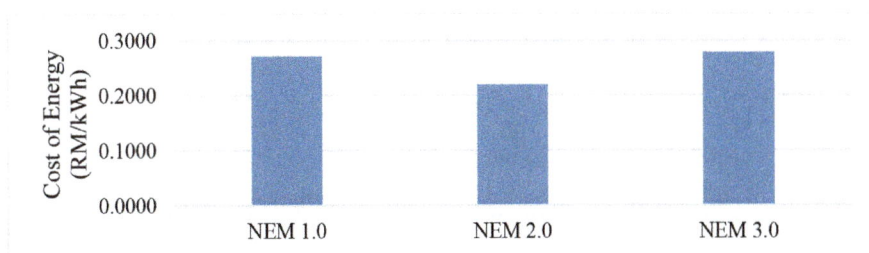

Figure 6: *COE* parameters for different NEM schemes.

system cannot export excess electricity into the grid. Thus, the consumer will have to pay the import bill. Using the import bill in Table 7, the O & M cost can be calculated for NEM 3.0 by adding the reduced bill for 10 years at about RM 22,700 and the import bill for another 10 years at RM 44,050, which produces the highest *NPC* of RM 100,742. From the *NPC*, the annual net savings can be obtained, and it is expected to have the highest savings from the lowest *NPC*. The annual net savings from NEM 1.0 and 3.0 are RM 1,787 and RM 1,664, respectively.

Since the *COE* value is also closely related to the *NPC*, the higher *COE* values in NEM 1.0 and 3.0 are expected, as shown in Fig. 6, with values of 0.2723 RM/kWh and 0.2792 RM/kWh, respectively. From the *COE* values, the NEM 3.0 still produces cheaper energy compared to the grid-only system.

5 CONCLUSION

This paper investigated the economic parameters of three different schemes of NEM implemented in Malaysia. The results showed that NEM 2.0 provides more benefits to the consumer compared to NEM 1.0 and 3.0 in terms of payback period, NPC, net savings and cost of energy. However, comparing the NEM schemes to the grid-only system, all schemes are capable of reducing electricity bills and generating acceptable annual net savings. This saving is greater if the lifetime of the inverter is longer. In addition, the actual bill calculation has also been explained to clarify misunderstandings with regard to NEM bills.

ACKNOWLEDGEMENTS

This research was supported by the Ministry of Higher Education Malaysia, through the Fundamental Research Grant Scheme (Ref: FRGS/1/2018/TK07/UMT/02/1). The authors would like to express their gratitude to the Faculty of Ocean Engineering Technology and Informatics, Universiti Malaysia Terengganu, for providing the facilities during the course of the research.

REFERENCES

[1] Blank, L. & Gegax, D., Do residential net metering customers pay their fair share of electricity costs? Evidence from New Mexico utilities. *Utilities Policy*, **61**, p. 100973, September, 2019. https://doi.org/10.1016/j.jup.2019.100973

[2] Razali, A.H., Abdullah, M.P., Hassan, M.Y., Said, D.M. & Hussin, F., Integration of time of use (TOU) tariff in net energy metering (NEM) scheme for electricity customers. *Indonesian Journal of Electrical Engineering & Informatics*, **7(2)**, pp. 255–262, 2019. https://doi.org/10.11591/ijeei.v7i2.1173

[3] Razali, A.H., Abdullah, M.P., Mat Said, D. & Hassan, M.Y., Annualized electricity cost of residential solar PV system under Malaysia's NEM scheme. *Elektrika - Journal of Electrical Engineering*, **19(1)**, pp. 50–54, 2020. https://doi.org/10.11113/elektrika.v19n1.204

[4] Desa, M.K.B.M., Masri, S. & Ganesan, L., An economic analysis for grid connected residential photovoltaic system in Malaysia. *Int. Conf. on High Voltage Engineering & Power Systems ICHVEPS 2017 - Proceeding*, pp. 145–148, 2017. https://doi.org/10.1109/ICHVEPS.2017.8225931

[5] Mansur, T.M.N.T., Baharudin, N.H. & Ali, R., Technical and economic analysis of net energy metering for residential house. *Indonesian Journal of Electrical Engineering & Computer Science*, **11(2)**, pp. 585–592, 2018. https://doi.org/10.11591/ijeecs.v11.i2.pp585-592

[6] "SEDA PORTAL." https://www3.seda.gov.my/. Accessed 03 March 2021.

[7] Razali, A.H., Abdullah, M.P., Hassan, M.Y. & Hussin, F., Comparison of new and previous net energy metering (NEM) scheme in Malaysia. *Elektrika - Journal of Electrical Engineering*, **18(1)**, pp. 36–42, 2019. https://doi.org/10.11113/elektrika.v18n1.141

[8] "Net Energy Metering (NEM) - FAQ - NEM Solar Malaysia." http://nemsolarmalaysia.com/nem-faq-solar-malaysia/. Accessed 03 March, 2021.

[9] Allouhi, A., Saadani, R., Kousksou, T., Saidur, R., Jamil, A. & Rahmoune, M., Grid-connected PV systems installed on institutional buildings: Technology comparison, energy analysis and economic performance. *Energy & Buildings*, **130**, pp. 188–201, 2016. https://doi.org/10.1016/j.enbuild.2016.08.054

[10] Li, H.X., Zhang, Y., Li, Y., Huang, J., Costin, G. & Zhang, P., Exploring payback-year based feed-in tariff mechanisms in Australia. *Energy Policy*, **150**, 112133, 2021. https://doi.org/10.1016/j.enpol.2021.112133

[11] Bhattacharjee, S. & Dey, A., Techno-economic performance evaluation of grid integrated PV-biomass hybrid power generation for rice mill. *Sustainable Energy Technologies & Assessments*, **7**, pp. 6–16, 2014. https://doi.org/10.1016/j.seta.2014.02.005

[12] Islam, M.S., A techno-economic feasibility analysis of hybrid renewable energy supply options for a grid-connected large office building in southeastern part of France. *Sustainable Cities & Society*, **38**, pp. 492–508, 2018, https://doi.org/10.1016/j.scs.2018.01.022

[13] Taghavifar, H. & Zomorodian, Z.S., Techno-economic viability of on grid micro-hybrid PV/wind/Gen system for an educational building in Iran. *Renewable & Sustainable Energy Reviews*, **143**, p. 110877, 2021. https://doi.org/10.1016/j.rser.2021.110877

[14] Garriga, S.M., Dabbagh, M. & Krarti, M., Optimal carbon-neutral retrofit of residential communities in Barcelona, Spain. *Energy & Buildings*, **208**, 2020. https://doi.org/10.1016/j.enbuild.2019.109651

[15] Maammeur, H., Hamidat, A., Loukarfi, L., Missoum, M., Abdeladim, K. & Nacer, T., Performance investigation of grid-connected PV systems for family farms: Case study of North-West of Algeria. *Renewable & Sustainable Energy Reviews*, **78**, pp. 1208–1220, 2017. https://doi.org/10.1016/j.rser.2017.05.004

[16] Mukisa, N., Zamora, R. & Tjing Lie, T., Store-on grid scheme model for grid-tied solar photovoltaic systems for industrial sector application: Costs analysis. *Sustainable Energy Technologies & Assessments*, **41**, p. 100797, 2020. https://doi.org/10.1016/j.seta.2020.100797

[17] S. E. D. A. M. SEDA, "Annual Report 2013," 2013.

[18] S. E. D. A. M. SEDA, "SEDA Annual Report 2017," 2017. [Online]. Available: http://www.seda.gov.my/download/seda-annual-report/

[19] "Pricing & Tariffs - Tenaga Nasional Berhad." https://www.tnb.com.my/residential/pricing-tariffs/ (accessed 08 August, 2021).

[20] S. E. D. A. M. SEDA, "SEDA Annual Report 2018," 2018. [Online]. Available: http://www.seda.gov.my/download/seda-annual-report/

THE CONSULTATIVE DRAFTING PROCESS FOR CHINESE EMISSIONS TRADING REGULATION: EFFECTIVE INPUTS BUT UNCERTAIN OUTPUT

STEVEN GEROE
Faculty of Law, La Trobe University, Melbourne, Australia.

ABSTRACT
The interviews informing this paper provide a series of interlocking case studies of the ways in which specialist expertise in renewable energy institutions is integrated through the consultative drafting processes for Chinese emission trading schemes (ETS). This has been implemented through drafting groups, research collaboration, various types of meetings and conferences, industry feedback and online solicitation of opinions. Interviews in state-related research institutions, universities, regional ETS carbon exchanges and private sector consultancies indicated that this process can be a useful means of integrating regulatory measures that have proven effective. Not all interviewee recommendations are reflected in the February 2021 Trial Measures for the Chinese national ETS and related implementing rules. Examples of adoption of interviewees' recommendations included detailed requirements for emissions monitoring plans, models for trading systems and registries, and specific methodologies such as default emissions values. The most important examples of non-adoption were stringent penalties for emissions and monitoring, reporting and verification (MRV) offences and supervisory powers of regulators relating to third-party inspection organisations. The March 2021 opinion solicitation draft (OSD) for a higher level, more permanent State Council regulation contains stronger penalties and supervisory powers. The 2019 and 2020 OSDs for the current national ETS rules also contained stronger penalties and supervisory powers than the current rules. Hence, all of the OSDs more closely resemble interviewee recommendations than the current rules. Interview evidence, and related scholarly writing, suggests that this pattern may relate to resistance of powerful economic interests. Nonetheless, it suggests that such resistance can be countered through capacity building and the example of early adopters in effective emissions trading. While the consultative drafting process has proved a useful means for identifying effective regulatory design on the basis of pilot ETS experience, it has yet to be put to best effect in the current national ETS rules.
Keywords: carbon price, Chinese, consultative drafting, emissions trading, regulation, renewable energy.

1 INTRODUCTION
The interviews informing this paper provide a series of interlocking case studies of the ways in which specialist expertise in renewable energy institutions is integrated through the consultative drafting processes for Chinese emission trading schemes (ETS). It does not purport to provide an account of all institutions and processes involved. Rather, it provides illustrative examples of means by which stakeholders interviewed provided input into the ETS drafting process on design features they regarded as critical. This addresses a gap in the research on stakeholder engagement in the drafting of the national ETS. As Stoerk *et al.* note, 'The academic literature on this point is scarce' [1].

The National Development and Reform Commission (NDRC) and the Ministry of Ecology and Environment (MEE) have in turn coordinated the input of a range of institutional stakeholders into the regulatory process. This has largely been implemented through drafting groups, research collaboration, various types of meetings and conferences, industry feedback and online solicitation of opinions. Interview responses indicated that the consultative drafting process can be a useful means of integrating regulatory measures that have proven effective. Research interviews were conducted in state-related research organisations, universities, carbon trading exchanges and private sector entities involved in ETS drafting in Beijing, Tianjin, Guangzhou and Wuhan, from December 2019 to January 2020. A semi-structured interview format was adopted. This involved follow-up questions exploring responses to general questions on integration of learning from the pilots into ETS rules through institutional involvement in consultative drafting. This enabled flexibility in pursuing examples of regulatory approaches within interviewees' own experience. The paper is structured to examine the contribution of different types of institution in the consultative drafting process for the regional pilot and the Chinese national ETS.

A number of interviewee recommendations are not reflected in the February 2021 Trial Measures for the Chinese national ETS and related implementing rules. Examples of adoption of interviewees' recommendations included detailed requirements for emissions monitoring plans, models for trading systems and registries, and specific methodologies such as default emissions values. The most important examples of non-adoption were stringent penalties for emissions and monitoring, reporting and verification (MRV) offences and supervisory powers of regulators relating to third-party inspection organisations. The March 2021 opinion solicitation draft (OSD) for a higher level, more permanent State Council regulation contains stronger penalties and supervisory powers. The 2019 and 2020 OSDs for the current national ETS rules also contained stronger penalties and supervisory powers than the current rules. Hence, all of the OSDs more closely resemble interviewee recommendations than the current rules. Interview evidence, and related scholarly writing, suggests this pattern may relate to resistance of powerful economic interests. Nonetheless, it suggests that such resistance can be countered through capacity building and the example of early adopters in effective emissions trading. While the consultative drafting process has proved a useful means for identifying effective regulatory design on the basis of pilot ETS experience, it has yet to be put to best effect in the current national ETS rules.

2 INTERVIEW EVIDENCE

2.1 Regional pilot ETS

The State Council has been responsible for the development of the regional pilots and the Chinese national ETS, as part of its administrative function of drawing up and implementing five-year plans for China's economic and social development [2]. As such, the national legislature, the National People's Congress (NPC), has not been directly involved in either regional or national ETS development. The seven regional ETS pilots were initiated by planning documents and rules of the NDRC [3], formerly known as the State Planning Commission. The regional pilots have been implemented by a combination of Local People's Congress (LPC) rules [3], [4] and [5], and rules of sub-national departments of the NDRC and subsequently the MEE. These sub-national rules have provided the implementation detail for the broad principles and objectives set out by the NDRC [6]. (The Chinese Legislation Law sets out a hierarchy of legislative authority, in descending order as follows: laws of the NPC, national administrative regulations of the State Council and rules of central government departments made to implement State Council regulations or decisions. Rules made by sub-national People's Congresses rank higher than rules of sub-national executive government ministries) [7].

With regard to the consultative drafting process of the regional pilots, Guangdong is taken as an example due to the detailed interview evidence provided. In 2011, the NDRC selected

Guangdong as one of seven pilot ETS. Several institutions participated in a working group led by the Guangdong Development and Reform Commission (the Guangdong DRC). These included the Guangdong Research Centre for Climate Change at Sun Yat-sen University, the Guangzhou Energy Research Centre of the Chinese Academy of Sciences, the Guangdong section of the Chinese Academy of Social Sciences, the China Quality Certification Centre and the Ministry of Industry and Information. Research topics for reports to the Guangdong DRC were allocated to each of these institutions. Interviewee Zeng Xuelan participated as Director of the Guangdong Research Centre for Climate Change at Sun Yat-sen University [8]. Three regulations were drafted: for MRV, quota allocation and management of the whole ETS (including trading). Specific annual allocation plans were also drafted.

The working group initially focused on MRV. As the emissions volume of the covered enterprises was unknown, the working group visited and studied the data and statistical systems of the more than three hundred covered enterprises and government institutions. This involved close cooperation with the relevant industry organisations. In the first phase, the power, iron and steel, and petrochemicals industries were covered, followed by the aviation and paper-making industries. Zeng Xuelan indicated that: 'Local industry organizations made an important contribution [to scheme design]. We sought their suggestions, at meetings and through written submissions. At the start our main focus was on researching Guangdong enterprises, and studying the EU and Californian ETS.' The working group also had close cooperation with experts on the EU and Californian ETS, through research meetings.

Requirements for emissions monitoring plans are specified in the Guangdong MRV rules [9]. Since 2018, Guangdong (unlike some pilots) has required monitoring plans, set deadlines and required information, such as monitoring boundaries of the company, monitoring methods and data selection [8]. Pilot experience shows that requiring adequate monitoring plans improves data quality, and monitoring and reporting of emissions [4] and [23]. Zeng Xuelan indicated that the most important lesson from experience of monitoring plans in Guangdong is that they should not be revised in the absence of important changes to the enterprise's situation. Otherwise, they should only be changed to make the requirements more stringent. As monitoring plans are drafted by covered entities, this requirement could prevent any weakening of monitoring requirements by entities without adequate justification. Under MRV guidelines for enterprises in the national ETS issued by the MEE in March 2021, Data Quality Control Plans can only be amended in specified circumstances. These include where new emissions are due to changes in facilities, fuels or materials; where new methods or instruments are adopted to improve data accuracy; and where existing monitoring methods are incorrect, can be improved or do not meet requirements [10]. Data Quality Control Plans must include, among other things, accounting boundaries and methods, production data, emissions factors and measures for internal data quality control. They must comply with all applicable MEE emissions accounting guidelines, and technical specifications and standards [10]. While these requirements are consistent with interviewees' recommendations, they are contained in an MEE departmental guideline without applicable penalties for non-compliance. The financial penalty for false emissions reporting in the 2021 Trail Measures Article 39 applies to GHG reports submitted to the MEE, not Data Quality Control Plans. The maximum penalty of 30,000 RMB in that provision is much lower than 200,000 RMB in both the 2019 OSD [11] and the March 2021 OSD for a State Council regulation [12].

Zeng Xuelan also referred to Guangdong's approach to supervision of emissions verification organisations. While under Beijing pilot rules, the MEE selects some third-party verifiers to check other inspection agencies' reports; in Guangdong, all third-party verifiers must be registered. All third-party verification reports are cross checked by other inspection organisations, as opposed to only selected reports being checked as in the other pilots. If mistakes are discovered, a further spot verification is conducted. If no mistakes are discovered, some results are sampled and subjected to spot checks on the enterprise. Additionally, an evaluation system for third-party verifiers involves a score based on evaluation rules. The score is based on the number of errors in the report and the occurrence of any illegal conduct such as conflict of interest. A third-party verification blacklist precludes third-party verifiers receiving work for the following 12 months [8].

Article 31 of the 2021 Trial Measures requires the MEE to implement verification work according to the 'double random one open' method, meaning random selection of inspection subjects and inspectors, and public disclosure of inspection outcomes [13]. It also empowers local MEE departments to determine the focus and frequency of inspections, on the basis of verification outcomes on liable entities' emissions reports. It does not refer to powers to make spot checks and to check and/or copy relevant documents and materials, or trading-related information systems and monitoring facilities in the manner of 2019 OSD Article 18. Article 22 of the 2021 OSD for a State Council regulation is stronger in this regard, as it provides that the MEE shall supervise and manage both trading entities and verification organisations through onsite inspections, copying, checking and investigating relevant documents, materials and information systems, and requiring explanations for any relevant issues. Additionally, the MEE shall establish a mechanism for sharing information and facilitating law enforcement with market, banking and securities regulatory authorities. Nonetheless, there is a need for subsequent regulations to provide penalties for the full range of misconduct specified by the 2021 MEE MRV rules. This includes accepting funding from Key Emissions Entities, participating in carbon asset management or carbon trading activities, sharing personnel with Key Emissions Entities being verified and using inspectors with a conflict of interest [10]. While a credible approach, the double random one open method is a less comprehensive means of supervising verification organisations than the measures adopted in Guangdong. Combining the supervisory measures in the Guangdong pilot with stronger MEE investigative powers, and financial penalties for all conduct prohibited by MEE rules for verification organisations, would provide a stronger regulatory basis for reliable MRV data.

State-related research institutions have also contributed to the development of the Tianjin pilot ETS. While Tianjin Academy of Environmental Science (TAES) research covers environmental issues more broadly, the Tianjin Low-carbon Research Centre (TLCRC) is more specifically focused on ETS research, particularly relating to MRV. For example, a methodology for emissions calculations for steelmaking based on industrial processes as opposed to financial factors was developed at the TLCRC for the Tianjin ETS, which will be adopted in the national ETS [6]. TLCRC is also engaged in research on other ETS design elements, such as permit allocation. TAES and TLCRC are both specialised research centres affiliated with the Tianjin bureau of the MEE. This research is complementary to that of Nankai University, which is responsible for the MRV system and regulations. The Tianjin Carbon Emissions Trading Exchange (TCX) is involved in development of market trading rules and related systems. The Tianjin University of Science and Technology provides expertise on measurement and quantification and other technical aspects of data provision relevant to MRV, permit allocation and other elements of ETS operation [6]. TAES and TLCRC participate in consultative drafting, through meetings organised by the MEE, typically when new policies are introduced. They can be held in Beijing or one of the pilot cities [6].

2.2 The national ETS

The MEE coordinates consultation with its specialist research institutions and with industry organisations on the ETS, through meetings and conferences. It has coordinated many rounds of consultation with ministries and government agencies involved in the ETS. These include the Ministry of Finance, the Ministry of Commerce, the Ministry of Industry and Information, the Ministry of Science and Technology, the China Civil Aviation Authority and the China Securities Regulatory Commission [14]. It also publishes OSD rules for the national ETS on its website [11] and [15]. Duan Maosheng indicated that the Institute of Energy, Environment and Economy (IEEE) at Tsinghua University was responsible for drafting the 'specific language' of the OSDs for the national ETS regulations. It reports to the MEE on whether suggestions should be included or taken into account in the revised rules. Tsinghua is the only university involved on an institutional basis in drafting the national ETS rules [14]. Individual academics from several universities including the China University of Politics and Law also contribute. The Guangdong Research Centre for Climate Change at Sun Yat-sen University has also supported the MEE through submissions on the OSDs for the national ETS, and indirectly by providing research and suggestions to the Guangdong provincial government, who pass them on to the MEE [8]. The IEEE has members from disciplines including engineering, energy, and finance and economics.

Duan Maosheng indicated that penalties imposed per t/CO_2e can be a stronger disincentive for excess emissions than one-off fines. The Beijing ETS imposes financial penalties set at three times the average market prices per t/CO_2e for excess emissions up to 10% of allocations, five times for excess of over 20% and four times for 11–20% [16]. He indicated this penalty was a factor for achieving high rates of compliance. This model was adopted in Article 19 of the 2019 OSD for the national ETS. It provided that, following expiration of a warning period, the local bureau of the MEE can impose a penalty 2–5 times the average market price per t/CO_2e for that year [11]. Article 40 of the 2021 ETS Trial Measures provides for one-off fines of 20,000–30,000 RMB. Where the shortfall is not corrected within a specified deadline, the amount of the shortfall shall be deducted from the following year's quota allocation [17]. Article 25 of the March 2021 OSD for a State Council national ETS regulation follows the same approach, but with substantially higher fines of 100,00–500,000 RMB. While the threat of a correspondingly tighter emissions allocation creates additional incentive for performance, it may be less stringent than a penalty applied per t/CO_2e for more significant shortfalls.

Carbon exchanges have played a significant role in drafting trading rules and administrative support systems such as registries, trading platforms and clearance mechanisms. For example, the China Hubei Emission Exchange has participated in consultative drafting through workshops on key issues for the national ETS, along with representatives of the other pilots, the big electricity companies, industry associations and universities. It is a company owned by the Hubei provincial government [18]. The President of the Hubei exchange also contacts the MEE directly, providing a further channel for involvement in regulatory development. The Hubei exchange proposal for design of the registry for the national ETS was selected for implementation by the MEE [19]. Based on the existing Hubei registry, it incorporated improvements based on lessons from implementation experience. Tian Yiran stated that the essential priorities are for the system to be easy to use and control and to provide security for carbon asset sales. For example, the existing coding system based on the EU ETS has a unique identifier for each discrete ton of GHG. The proposed model will have a common code for all tons of GHG from the same source/entity. This will be more efficient and will reduce the burden on the IT and administrative system. Special accounts can be created for related companies, to manage carbon assets on the registry in one account. Fields can be created for separate entries to differentiate between parties with different roles in carbon asset management in the same entity, such as personnel responsible for making and reviewing applications. This capability also facilitates development of enterprises' carbon asset management systems 18].

The China Beijing Environmental Exchange (CBEX) is also a state-owned company, owned by the Beijing municipal government. It prepares reports for the MEE on the performance of the Beijing pilot, available on its website [20]. The Guangdong Climate Exchange and the TCX are also involved in consulting on ETS trading rules [6, 8]. The Shanghai pilot proposal for the trading system for carbon units has been selected for the national ETS [21]. Additionally, all carbon exchanges conducted capacity building programs to support compliance by covered entities [22]. Yu Zexia said that:

'In Hubei, in the first year of the pilot, many companies were angry about the ETS regulation and not afraid to show it. They refused to comply or buy allowances or register accounts at the exchange. We encouraged the companies to participate. In 2014 we held seven capacity building sessions, training companies how to participate. When some companies made some revenue under the trading system others became more willing to participate.'

In this way, capacity building programs implemented by the exchanges were an important element in ensuring participation and compliance by covered entities. This role, together with tendering to the MEE to design operational elements of the national ETS, is additional methods of contributing to ETS development, alongside collaborative research, meetings and direct communication with government.

SinoCarbon Innovation and Investment Ltd. is a private sector entity mentioned by a number of interviewees as closely involved with the consultative drafting process for the national ETS [8, 18, 20]. It is a consulting company and also provides third-party emissions verification services. It does not engage in carbon trading or carbon asset management. It produces software used by firms in carbon accounting and trading, MRV and data management. It works with both government and business, and it is involved in projects supporting provincial governments to prepare for the national ETS and on MRV guidelines for specific sectors [22]. It has been involved in cooperative projects with the World Bank, the Asian Development Bank and the international consulting company ICF. It has received EU funding for projects to support policy dialogue, involving communication between policy makers from the EU and China, as well as capacity building for local officials, MEE officials and industry executives. These projects have covered topics including MRV training of technical staff and trading simulation. Such EU-China joint capacity building projects have been implemented in all provinces. Other project partners of SinoCarbon include Tsinghua University, the National Climate Strategy Center, the World Bank Partnership for Market Readiness and the International Carbon Action Partnership (ICAP). SinoCarbon provides an annual report to ICAP on the Chinese ETS and contributes to ICAP international carbon market reports. SinoCarbon is involved in consulting for the MEE with regard to the rules for the national ETS, e.g. offset rules [22].

Similarly to Zeng Xuelan, Guo Wei of SinoCarbon recommends legislative requirements for monitoring plans covering 'monitoring boundaries of the company (the source streams, emission sources, activities, etc.), the monitoring methodology (e.g. default values, sampling standards), methods used to determine the different parameters such as emission factors, description of the quality assurance and the quality control system, included in uniform national templates for monitoring plans [4, 22. Similarly to Duan Maosheng, he referred to the inadequacy of some one-off fines as a deterrent for excess emission, as in Shanghai, where the level of the fine was on average lower than the marginal cost of abatement for emissions.

The MEE also consults directly with the China Electricity Commission (CEC), regarding allocation methods under the OSD process [22]. Other industry groups such as power and steel organisations are also consulted directly or through workshops and meetings [14]. For example, the China Huadian Corporation, one of the five large state-owned power companies in China, helped to develop a low carbon plan to minimise the impacts of national ETS [22]. The MEE also seeks written responses from industry groups and local governments on ETS design issues [14].

3 CONCLUSIONS

Interview evidence suggests that the consultative drafting process is an effective means of integrating specialist expertise into ETS regulatory development. Nonetheless, the input of specialist expertise is clearly not a guarantee of its inclusion in ETS rules. In some cases, such as the development of guidelines for monitoring plans, regulatory output is quite consistent with interviewees' recommendations. Conversely, the level of penalties for emissions and MRV offences in current national ETS rules is less stringent than interviewee recommendations, and provisions for supervision of third-party verification organisations are less rigorous. In these regards, the provisions of the 2021 Trial Measures are less stringent than those in the 2019 and 2020 OSDs for MEE rules, and the 2021 OSD for a future State Council regulation. This may reflect resistance to ETS implementation from some stakeholders. As Duan *et al.* observe: 'During the starting period of China's national ETS, policy makers must fully consider the possibility of enterprises underestimating the government's determination to enforce ETS, large enterprises being united to reject responsibilities, enterprises misunderstanding the role of ETS and lacking awareness of carbon asset management, and reluctant sellers in the allowance market' [23]. Consistently with Hubei interviewees, they emphasise capacity building as a primary means of overcoming such resistance. Commercial stakeholders have not been the only parties resistant to national ETS implementation, with some local government officials 'concerned that the national system will worsen the economic situation and thus be resistant to this policy' [sic] [24]. While the consultative drafting process has proved a useful means for identifying effective regulatory design on the basis of pilot ETS experience, it has yet to be put to best effect in the current national ETS rules. That will require addressing stakeholder resistance through capacity building, familiarisation through successful emissions trading, and commitment to sufficiently stringent measures in a higher level, more permanent State Council regulation.

REFERENCES

[1] Stoerk, T., Dudek, D.J. & Yang, J., China's national carbon emissions trading scheme: lessons from the pilot emission trading schemes, academic literature, and known policy details. *Climate Policy*, **19(4)**, pp. 472–486, 2019. https://doi.org/10.1080/14693062.2019.1568959

[2] Constitution of the People's Republic of China [中华人民共和国宪法], National People's Congress (adopted and effective on 4 December 1982; amended on 12 April 1988, 29 March 1993, 15 March 1999, 14 March 2004 and 11 March 2018).

[3] Guidelines for Accounting and Reporting GHG Emissions for Key Enterprises; NDRC 2013, 2014 and 2015; Interim Measures for the Administration of Carbon Emissions Trading, NDRC 2014; Notice on Implementation of Activities to set up Carbon Emissions Trading Market, NDRC 2016.

[4] Tang, R., Guo, W., Oudenes, M., Li, P. & Tang, J., Key challenges for the establishment of the monitoring, reporting and verification (MRV) system in China's national carbon emissions trading market. *Climate Policy*, **18**, Supplement 1, pp. 106–121, 2018. https://doi.org/10.1080/14693062.2018.1454882

[5] Peng, F. & Yan, L., The institutional comparison research of 7 pilot ETS-based on the review on the legal documents of the 7 pilots. *China Environment Law*, **2**, pp. 25–45 2014.

[6] Interview with Wang Dai, Li Zhou and Zhang Lin, offices of the Tianjin bureau of the MEE, 16 December 2019. Wang Dai is a member of the Tianjin Academy of Environmental sciences. Li Zhou and Zhang Lin are members of the Tianjin Low-Carbon Research Centre.

[7] Legislation Law of the People's Republic of China [中华人民共和国立法法], National People's Congress (adopted on 15 March 2000, effective on 1 July 2000 and amended on 15 March 2015), Articles 7, 8, 11, 56, 71 and 73.

[8] Interview with Zeng Xuelan and Li Weichi, Sun Yatsen University, 4 January 2020. The Low-Carbon Research Centre of Sun Yat-sen University was the forerunner of the Guangdong Research Centre for Climate Change.

[9] Guangdong Development and Reform Commission (DRC) Notice on Implementation Measures on Monitoring, Reporting and Verification of Carbon Emissions for Enterprises in Guangdong (Trial) [广东省发展改革委关于印发《广东省企业碳排放信息报告与核查实施细则（试行）》的通知], Guangdong DRC Order No. (2014) 145, 18 March 2014.

[10] Notice on Enterprise GHG Emissions Reporting Verification Guide (trial) 关于印发《企业温室气体排放报告核查指南（试行）》的通知, MEE 26 March 20201, [4.2.1.6.1 a)], 2.3 and [4.1.1], with detail at 4.2.1.1 - 4.2.1.7. http://www.mee.gov.cn/xxgk2018/xxgk/xxgk06/202103/t20210329_826480.html

[11] MEE, 'Carbon Emissions Trading Management Interim Regulations (Opinion Solicitation Draft)' [碳排放权交易管理暂行条例（征求意见稿）] 3 April 2019, available at: https://www.mee.gov.cn/hdjl/yjzj/wqzj_1/201904/t20190403_698483.shtml

[12] Interim Regulations on the Administration of Carbon Emissions Trading (Revised Draft) 碳排放权交易管理暂行条例（草案修改稿）http://www.mee.gov.cn/xxgk2018/xxgk/xxgk06/202103/W020210330371577301435.pdf English translation http://www.cet.net.cn/uploads/soft/202104/1_15100052.pdf

[13] International Carbon Action Partnership, China publishes two major policy draft for national ETS, November 2020. https://icapcarbonaction.com/en/news-archive/728-china-publishes-two-major-policy-drafts-for-national-ets

[14] Interview with Duan Maosheng, Tsinghua University Beijing, 20 December 2019. Duan Maosheng is the Director of the Institute of Energy, Environment and Economy, Tsinghua University; and Director, China Carbon Market Centre, Tsinghua University.

[15] International Carbon Action Partnership, China Releases Draft Interim Regulations on the Management of Carbon Emissions Trading. https://icapcarbonaction.com/en/news-archive/628-china-releases-draft-interim-regulations-on-the-management-of-carbon-emissions-trading

[16] Zhang, Z., Carbon Emissions Trading in China: The Evolution from Pilots to a Nation-wide Scheme, Working Paper 1503, Crawford School of Public Policy Centre for Climate Economics& Policy, Australian National University, 14, 2015. Available at: https://ccep.crawford.anu.edu.au/publication/ccep-working-paper/5535/carbon-emis-sions-trading-china-evolution-pilots-nationwide

[17] Ministry of Ecology and Environment (MEE), Carbon Emissions Trading Management Measures (Trial) [碳排放权交易管理办法（试行）], MEE Order No. (2021) 16, 31 December 2020 (effective 1 February 2021).

[18] Interview with Yu Zexia and Tian Yiran at Hubei Carbon Exchange, Wuhan, 6 January 2020. Yu Zexia is Manager and Tian Yiran is Vice Manger of the trading department, Hubei Carbon Exchange.

[19] International Carbon Action Partnership (ICAP) ETS Detailed Information: China, Hubei', https://icapcarbonaction.com/en/?option=com_etsmap&task=export&format= pdf&layout=list&systems%5B%5D=58

[20] Interview with Yan Lei and Professor Liu Yonggong, China Agricultural University, Beijing, 19 December 2019. Yan Lei is a Senior Manager in the Carbon Trading Center of the China Beijing Environmental Exchange.

[21] Hua, Y. & Dong, F., China's carbon market development and carbon market connection: A literature review. *Energies*, **12(9)**, p. 1663, 2019. https://doi.org/10.3390/en12091663

[22] Interview with Guo Wei, Chen Zhibin and Jia Hui, offices of Sino Carbon, Beijing, 18 December 2019.

[23] Deng, Z., Li, D., Pang, T. & Duan, M., Effectiveness of pilot carbon emissions trading systems in China. *Climate Policy*, **18(8)**, pp. 992–1011, 2018. https://doi.org/10.1080/14693062.2018.1438245

[24] Zeng X., Duan, M., Yu, Z., Li, W., Li, M. & Liang, X., Data-related challenges and solutions in building China's national carbon emissions trading scheme. *Climate Policy*, **18**, sup1, pp. 90–105, 2018. https://doi.org/10.1080/14693062.2018.1473239

Author index

WIT*PRESS* ...for scientists by scientists

Energy and Sustainability IX

S. SYNGELLAKIS, *Wessex Institute, UK*

The world's economy is fuelled by energy. Depletion of resources and severe environmental effects resulting from the continuous use of fossil fuels has motivated an increasing amount of interest in renewable energy resources and the search for sustainable energy policies. This volume contains research papers presented at the 9th International conference on Energy and Sustainability.

The changes required to progress from an economy mainly focussed on hydrocarbons to one taking advantage of sustainable renewable energy resources require considerable scientific research, as well as the development of new engineering systems. Energy policies and management are of primary importance to achieve the development of sustainability and need to be consistent with recent advances in energy production and distribution.

In many cases, the challenges lie as much in the conversion from renewable energies (wind, solar, etc.) to useful forms (electricity, heat, fuel) at an acceptable cost including damage to the environment as in the integration of these resources into the existing infrastructure.

The diverse topics covered by the papers in this book involve collaboration between different disciplines in order to arrive at optimum solutions, including studies of materials, energy networks, new energy resources, storage solutions, waste to energy systems, smart grids and many others.

These research papers put a focus on sustainability across the multidisciplinary components of urban planning, the challenges presented by the increasing size of cities, the number of resources required and the complexity of modern society.

WIT Transactions on Ecology and the Environment, Vol. 254
ISBN: 978-1-78466-449-7   eISBN: 978-1-78466-450-3
Published 2022 / 194pp

www.ingramcontent.com/pod-product-compliance
Lightning Source LLC
Chambersburg PA
CBHW062008190326
41458CB00009B/3013